CZECHOSLOVAK ACADEMY OF SCIENCES

ENTROPY *and* INFORMATION *in Science and Philosophy*

CZECHOSLOVAK ACADEMY OF SCIENCES

Scientific Editor: Academician Jaroslav Kožešník
Reviewer: Karel Winkelbauer

ENTROPY *and* INFORMATION *in Science and Philosophy*

edited by LIBOR KUBÁT *and* JIŘÍ ZEMAN

ELSEVIER SCIENTIFIC PUBLISHING COMPANY
AMSTERDAM – OXFORD – NEW YORK 1975

Published in co-edition with
ACADEMIA, Publishing House of the Czechoslovak Academy
of Sciences, Prague

Distribution of this book is being handled by the following publishers

for the U.S.A. and Canada

American Elsevier Publishing Company, Inc.
52 Vanderbilt Avenue
New York, New York 10017

for the East European Countries, China, North Korea, Cuba, North Vietnam and Mongolia

Academia, Publishing House of the Czechoslovak Academy of Sciences, Prague

for all remaining areas

Elsevier Scientific Publishing Company,
335 Jan van Galenstraat
P. O. Box 211, Amsterdam, The Netherlands

ISBN 0-444-99888-8
Library of Congress Card Number 73-90961

Printed in Czechoslovakia

PREFACE

The volume "Entropy and Information" issues in certain sense from the aim and ideas of the previous volume "Time in Science and Philosophy" edited by J. Zeman (Academia and Elsevier, Prague – Amsterdam 1971): to collect papers written by leading scientists from various scientific branches and of different countries devoted to the general problems of an important interdisciplinary concept. It was the concept of time which made the centre of interest in the first volume and the object of seeking for a common language and unifying topic in science.

The present volume choses the concepts of entropy and information as the field of interest. Entropy was developed in physics in the 19th century and was applied later in astronomy, chemistry and biology. The rise of cybernetics and information theory brought to this concept a new and broader sense. Entropy, closely connected with the concept of information, influenced in its mathematical and philosophical form almost every science. It is of course well known that there are many various views on this concept. Some scientists think, for instance, that the concepts of entropy in physics and in information theory are quite different. Nevertheless, we feel it is fruitful to present new issues concerning the entropy concept from different viewpoints, to expose the problems and to endeavour in this manner to achieve a conceptual connection.

The volume is divided into three parts. The first part deals with the concept of entropy and information in the physical sciences, the second part concerns these concepts in life sciences and the third part studies the general and methodological problems of entropy and information. The first part investigates the entropy problem in the universe (Davies, Uyomov), in physics (Harrison), in relativity theory (Costa de Beauregard, Schlegel), in information theory (Peters), and in system theory (Denbigh). In the second part the biological and bio-economic problems of entropy and information are studied (Trincher, Georgescu-Roegen), as well as the neurophysiological (Braines – Svetchinsky) and the psychological questions (Strížcnec, Linhart – Vohník – Horák). The last part investigates the problem of the objectivity of entropy and information (Grünbaum, Ursul), the epistemological problems (Hörz, Zeman), and the ontological and general philosophical problems (Gal-Or, Szumilewicz).

We hope the reader will find in the volume some stimulating viewpoints although it was not possible to achieve in this book a quite uniform and total view.

L. Kubát *J. Zeman*

CONTENTS

ENTROPY AND INFORMATION IN THE PHYSICAL SCIENCES

I

COSMOLOGICAL ASPECTS OF IRREVERSIBILITY*

P. C. W. Davies

Time as a Fourth Dimension

We shall begin our discussion by regarding the world as an empty four dimensional manifold. For the moment we shall not worry about the global topology of the manifold, but concentrate instead on its local structure. Let us introduce in some small flat region test particles and fields. We may then define a local dynamics in accordance with the laws of special relativity, and we are led to construct a metric which is *pseudo*-Euclidean; that is, one dimension is distinguished from the other three by the opposite sign appearing in the metric. We call this special dimension "time". Null segments of the metric define the light cone, and all test particles move within the light cone on time-like world lines. It is assumed that the reader is familiar with these elementary aspects of special relativity.

At any point on a particle world line we may draw a tangential arrow which is also time-like (see Figure 1). This arrow may point either way, but suppose that we fix by convention a parti-

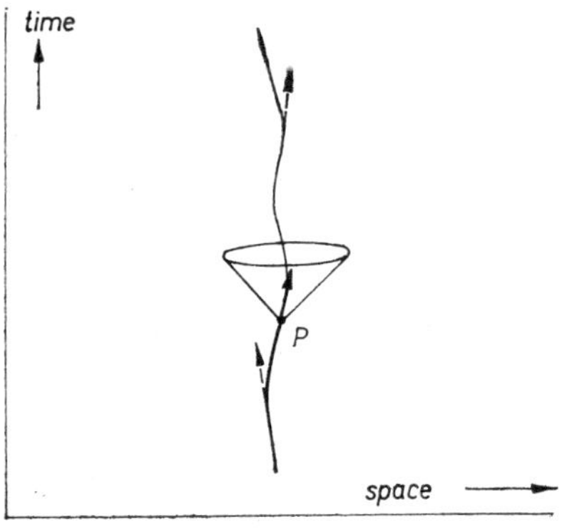

Fig. 1. The arrow at *P* will always point in the same direction if the world line is everywhere timelike and open.

* Much of this work was carried out while a Visiting Fellow at the Institute of Theoretical Astronomy, University of Cambridge.

cular direction. Then, locally at least,[1]) this direction is unchanged as we move the point P along the world line either way. Moreover, because of the structure of the Poincaré group, this direction is absolute in the sense of being the same for all observers.

At this stage all known elementary laws of physics (except those involving K mesons) in the region do not introduce a structural difference between the two directions of time. All such laws are invariant under time reversal, and the direction of our arrow is still purely a convention with no physical content.

We proceed to the subject matter of this article when we recognise that a structural difference does exist between the two directions of the time axis. The distinction between the past and future is very apparent in almost all local everyday phenomena. This is expressed by saying that most macroscopic systems encountered behave *irreversibly*. It is the purpose of this article to examine the connection between the local time asymmetry (irreversibility) of these phenomena and the global anisotropy of the manifold.

Entropy and Irreversibility

Most discussion on irreversibility starts with the second law of thermodynamics. This law states that the total entropy of a closed system is a monotonic function of time. In physical terms this means that if we encounter a closed system in non-equilibrium on a given occasion, then it will be close to equilibrium on another occasion sufficiently temporally separated from the first in one direction of time only. If this temporal direction between the two occasions is always the same for every closed system at all times (at least in our local space-time region), the law then endows a pervasive structural distinction between the two time directions, and we could unambiguously make the convention of calling the second (equilibrium) occasion the *later* occasion and affirm that entropy always *increases* with time (in a closed system) in our local region.

Boltzmann was able to derive this law from statistical mechanics in the limiting case of large systems, and provided a definition of the entropy in terms of the statistical weights of the macrostates.[2]) The law became degraded to a statement about probabilities; that in a large enough system the above entropic increase was overwhelmingly probable.

The fallaciousness of the statistical statement of the second law was exposed by Loschmidt, Zermelo and Poincaré, who rightly questioned Boltzmann's derivation of *irreversible* behaviour from fundamentally *reversible* laws of mechanics. Suppose that we consider an isolated box of gas. Every individual collision between two molecules is reversible in the sense that its time inverse is an equally acceptable dynamical solution of the equations of motion. Therefore, for

[1]) Cosmological models such as that of Gödel, which admit the existence of closed time-like world lines, will be excluded from our discussion here.

[2]) The interesting, though separate, philosophical problem of why entropy changes only occur in a *macroscopic* context i.e. by rejecting microscopic information in some sort of coarse-grained statistics, will not be discussed here.

every set of collisions that result in an entropy increase, there is a corresponding set that lead to a decrease, constructed by merely reversing all the molecular velocities. Poincaré formulated the objection in his periodicity theorem, which demonstrated how in a closed system over a very long period of time, the entropy will decrease as often as it will increase, consisting of a series of fluctuations about its maximum (equilibrium) value. Any given non-equilibrium state will recur with a frequency inversely correlated with the departure of that state from equilibrium, so that if the entropy increases with time from a given state on a number of occasions, then there must be an equal number of occasions when it will decrease to that value. The time extended behaviour of the entropy is therefore of the form shown in Fig. 2.

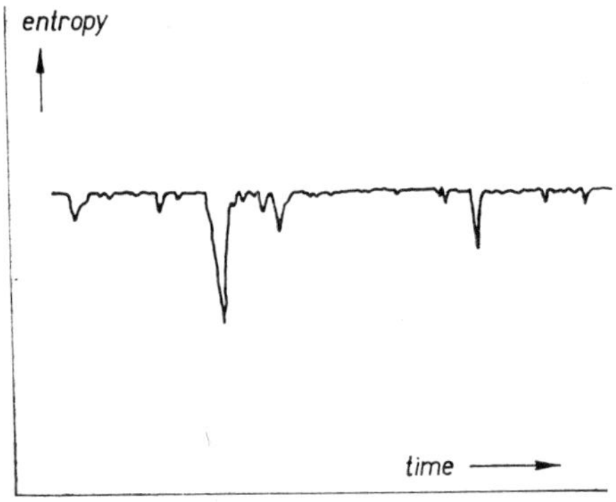

Fig. 2. Temporal behaviour of entropy in a closed system.

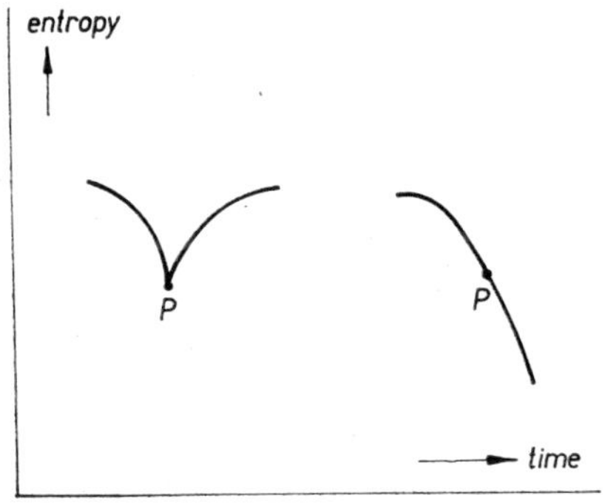

Fig. 3. The low entropy point P is far more likely to be symmetrically situated at a minimum rather than on a slope of a larger fluctuation.

There is no incompatibility of this type of objection with the statement that *if* the entropy is at a value less than maximum at a given time, then it is overwhelmingly probable that it will be at a higher value at a later time, provided we realise that it is also overwhelmingly probable that it was at a higher value at an *earlier* time. We can understand this by appreciating that should we encounter the system with an entropy value less than the maximum, it is far more likely that this value represents a minimum in a small fluctuation rather than the start (or finish) of a much larger one. This is because there are many more smaller than larger fluctuations (see Fig. 3). Moreover if the system is found in equilibrium at a given time, it is very probably far from one of the dips shown in Fig. 2, i.e., it will also be in equilibrium at nearby earlier and later times. In this way the overall time symmetry of the system is preserved and the entropic behaviour of an isolated closed system *cannot* provide an asymmetry of time after all.

It is here that we encounter a paradox. We are apparently to assume that the present entropic state of the world is far more likely to be a gigantic fluctuation from equilibrium, than to have evolved from a still lower entropy state in the past. How is it then that when we encounter a quasi-closed local system in a low entropy state, we invariably intuitively infer that this system has developed from a still lower entropy past state, or at least has remained undisturbed? As an illustration consider an example given by Grünbaum [1]. Suppose that while walking along a lonely beach we encounter a half obscured footprint in the sand. Subsequent patient observation will gradually reveal a slow increase in entropy of that region of the beach as the wind and tide erode the shape away. It is clear to us that entropy of the beach will increase in this fashion, but it is equally clear that prior to the encounter the footprint was sharper, i.e. in a still *lower* entropy state. How can this be reconciled with the above mentioned requirement that the encountered indentation be almost certainly at an entropy *minimum*, so that is overwhelmingly more likely to have formed prior to the encounter from a *higher* entropy state, with the chance arrangement of sand particles by the wind and tide?

The resolution of this paradox is available when we appreciate the concept of branch systems, originally discussed by Reichenbach [2]. Branch systems are subsystems of a larger system which separate off from the larger system to become quasi-closed for a while, after which they merge again with the larger system. Such branch systems can then be formed with a relatively lower entropy when they are coupled to the larger system, which itself suffers a more than compensatory increase in entropy. The arguments of Loschmidt and Poincaré are then no longer applicable, because the particular branch system simply did not exist as a quasi-closed system prior to the entropy minimum at which it separated out from the larger environment. Thus, in the case of the footprint, the slow deterioration by wind and tide was *not* mirrored prior to its discovery because that part of the beach was not a quasi-closed system when the creator of the footprint trod on the sand. At that time the system was open to the outside world, which certainly suffered an overall entropy increase as a result of the rearrangement. It must be emphasised however that we could not make these deductions if we knew nothing about feet, or the locomotive habits of humans.

As a second example suppose that we encounter a pea near the rim of a hemispherical bowl. We can be quite sure that after a few moments the pea will be at rest at the centre of the bowl, its potential and kinetic energy dissipated by entropy generating frictional processes. We could be equally sure that the pea was *not* at rest in the centre of the bowl a few moments *before* the encounter, but was probably dropped or fell into the bowl in the recent past. That is, we do not have to deduce that the chance arrangement of vast numbers of molecular collisions just happened to impart kinetic energy to the pea which took it to its position near the rim. We can conclude instead that the quasi-closed system pea + bowl only came into being as such just before the encounter.

Our world abounds with systems of this sort, naturally occuring as well as artificial, and our whole ability to interpret traces and records of the past rests on the above arguments. The resolution of the paradox hinges on the crucial distinction between open and closed systems which allows the formation of apparently entropy decreasing branch systems in an otherwise entropy increasing world. In this way structure and information accumulate in some regions as a result of their overall degradation in the whole system.

We must now inquire whether or not all branch systems undergo entropic changes in the *same* time direction. This is immediately apparent if we assume that an ensemble of branch systems are chosen in their initial states *at random*, which would certainly seem to be the case in practice. Referring to Fig. 2, which represents the time ensemble of a single closed system, we see that a randomly chosen state is either in equilibrium, or at an entropy minimum, with overwhelming probability. Therefore, a space ensemble with initial states randomly chosen from Fig. 2 would almost never include a branch system which *decreased* its entropy *relative to the others*. This argument, due to Grünbaum [1], furnishes us with an unambiguous operational definition of the "direction"[3]) of time at a probablistic level, as being the direction of increase of entropy of the vast majority of randomly formed branch systems. The importance of choosing more than one branch system in any satisfactory definition of the direction of time was originally elucidated by Schrödinger [3].

The above definition, which involves branch systems forming in low entropy states, is obviously meaningless unless the universe is in some thermodynamic non-equilibrium condition anyway. We must therefore face the question of whether or not we can discuss the entropic behaviour of the universe as a whole. For a finite universe it may be possible to define the total entropy and to apply a global second law of thermodynamics. Now we know that the formation of low entropy branch systems results in an overall increase in entropy of the total system. Consequently, such systems will only form in such a low entropy state, and subsequently exhibit their temporal entropic increases, in an outside world which itself is undergoing an overall entropic increase. In a closed world which is *decreasing* its entropy these branch systems would not therefore be formed in initially *random* states. They would instead be selected at the edge of dips in Fig. 2, and merge again into the outside world whenever they reached their minima. The basic time symmetry of a closed finite world is thus maintained.

Aside from questions of general relativity it is clear that we cannot define the total entropy, and hence the increase or decrease thereof, in an open infinite world model, because the total entropy then also becomes infinite. As such a world has an infinite Poincaré recurrence time we need not worry about overall time symmetry in the whole cosmological system. Neither can we define a direction of time from cosmological thermodynamics, contenting ourselves with our branch system definition instead.[4])

The Beginning of the World

When branch systems form by feeding on the negative entropy of a larger system, and possibly on the universe as a whole, it raises the curious question of where this negative entropy came from originally. Why is our world not in thermal equilibrium?

We shall mention three answers to this question. The first, a rather evasive answer, is the reply that if it were in equilibrium, we would not be alive to know about it. In fact Boltzmann even suggested that the present structure of the universe is due to a gigantic fluctuation from equilibrium, itself a precondition for the existence of biological matter. However, it is hard

[3]) The word "direction" is used here at a popular level. There is no connection between the discussion here and the apparent one way forward flow of conciousness in time, we being merely concerned with a physical distinction between opposite directions of the time axis. The direction is chosen purely by convention.

[4]) There may be other objections to defining the total entropy of the universe, such as the existence of horizons across which interactions cannot propagate.

to see why an exceedingly more probable fluctuation on the scale of the solar system alone would not suffice.

A second, and better, reply is that the universe is temporally finite, or at least semi-infinite. That is, the world was simply created a finite time ago in a low entropy state, and is running out its course to equilibrium, like a colossal branch system.

The third possibility (which has become very popular in recent years) is to appeal to a global temporal asymmetry – the expansion of the universe. We may obtain an understanding of this by a few simple considerations. In a homogeneous, isotropic, uniformally expanding universe we may define a unique velocity vector a teach point on the manifold. This is the velocity of the rest frame with respect to the average distribution of distant matter, i.e., the frame for which in all directions matter is receding at equal velocities at a given distance. It is also the frame in which a homogeneous distribution of black-body radiation appears isotropic. The collection of all these frames at every point is called the substratum. The proper time of a test particle at rest in the expanding substratum defines a unique priveliged time t, which by uniformity is the same for all such test particles over the whole substratum. We thus have a universal cosmic time t, and we may describe the expansion of the universe in terms of a single parameter $R(t)$ which plays the role of a length scale factor. At this epoch, and perhaps for all time, $R(t)$ is a monotonic increasing function of t, so defining a "direction" of time cosmologically. That is, establishing a *global* structural difference between the past and future directions of time on the manifold. The hypothesis is that the expansion of the universe generates entropy in some way. Superficially this appears a reasonable assumption. The expansion time scale $R/\dot{R}$ is far shorter than many important astrophysical relaxation times, so that the expansion is by no means quasi-static. According to laboratory thermodynamics, adiabatic expansion at a finite rate allows entropy to be generated. However, in the cosmological case the situation is somewhat different.

The first point to appreciate is that the finite expansion rate $\dot{R}/R$ does not in itself necessarily imply entropy generation, as was pointed out by Tolman many years ago [4]. For instance, consider a model universe uniformally filled with extreme relativistic gas, such as electromagnetic radiation. As the universe expands, the wave-length of each photon will scale with $R(t)$ i.e., the frequency (energy) ω will obey the relation

$$\omega \propto R^{-1}(t). \tag{1}$$

This is the basis of the well known apparent magnitude – red shift formula in cosmology. The expansion clearly does not change the distribution of photons among the various frequencies, because their wavelengths all scale in the same ratio. Consequently, if the radiation is thermal black-body at a temperature T_r, the black-body spectrum will be preserved under the expansion, though the temperature will decrease in proportion to the average photon energy:

$$T_r \propto R^{-1}(t). \tag{2}$$

The entropy density S is proportional to T_r^3, so that we have from (2)

$$SR^3 = \text{constant} \tag{3}$$

which says that the entropy in a comoving volume (one which expands with the universe) is constant in time, and so unaffected by the expansion. The system is therefore perfectly reversible. Although an observer in a fixed volume will see radiation flowing apparently irreversibly at a finite space, a reversal of the expansion back to a given value of R will simply return the radiation again and produce the *same* temperature T_r.

There is a similar argument in the case of a universe filled only with a non-relativistic monatomic ideal gas. A rescaling of lengths cannot affect the velocity distribution of the gas. In the non-relativistic case this is the same as the momentum distribution, so thermal equilibrium will be preserved under the expansion even at a finite rate. In a time dt an atom with velocity v relative to the substratum will travel a distance $v dt$. At this new point the substratum will be moving at a rate $\dot{R}/R\,(v\,dt)$ owing to the expansion, so that the measured velocity by a comoving observer at this point will be reduced by an amount $dv = -\dot{R}/R(v\,dt)$. Integrating this equation leads to the result $v \propto R^{-1}(t)$. Now the temperature, T_m, of an ideal gas is proportional to the average kinetic energy i.e. $T_m \propto v^2$, so that

$$T_m \propto R^{-2}(t). \tag{4}$$

Comparison of (2) and (4) shows that a non-relativistic gas cools faster than a relativistic one. Finally, the entropy in a volume R^3 is proportional to ln $(R^3 T_m^{3/2})$ in the monatomic case, so that using equation (4) it is seen that the total entropy in R^3 is constant in this case also. Note that this result does not depend on the collision rate in the gas, so the expansion need not be quasi-static.

From the point of view of conventional thermodynamics these results are somewhat surprising. Usually an adiabatic expansion can only be isentropic if it is also quasi-static. For example, in the case of a gas enclosed between a piston and cylinder a finite expansion rate will lead to turbulance, temperature gradients and friction between moving parts. Moreover, the finite relaxation time of the gas implies that the pressure behind the piston as it is withdrawn is less than when it is inserted, because of the time required for the gas to readjust to the increasing volume. All these effects preclude true reversibility and generate entropy, but all are absent in the cosmological case. There is no container with moving parts, and the uniformity throughout precludes temperature gradients and turbulence, and renders the relaxation time irrelevant. In this way the exact reversibility of our simple cosmological model can be made physically intelligible.

If we now consider a model universe uniformally filled with *both* matter and radiation coupled together then the above considerations are modified. This is apparent from relations (2) and (4) which indicate that in the absence of coupling a temperature difference will develop between matter and radiation. The equalisation of these temperatures would generate entropy. The system is no longer reversible, and entropy will also be generated during a contraction. However, if the coupling is strong, the equalisation rate is very much faster than the expansion rate and the process becomes more or less reversible. This will occur if the matter is strongly ionised, a situation which requires a temperature in excess of 5 000 °K. Referring to the relation (4) this will clearly occur for small enough R (or t). In practice the thermal capacity of the radiation vastly exceeds that of the matter in our universe (there are about 10^8 photons for each baryon). Consequently, before the decoupling at 5 000 °K the ions tend to follow the radiation temperature relation (2) by absorbing photons. After recombination the decoupled atoms soon cool to a negligable temperature. There are however some theoretical and observational reasons to conjecture that intergalactic space may contain large quantities of re-ionised hydrogen. Although this particular type of entropy generating process is inconspicuous and probably completely negligable, it is nevertheless interesting because it depends directly on the continued cosmological expansion (or contraction). Layzer has made it the fundamental origin of irreversibility [5]. There are other processes of this variety; for example, ionic and molecular recombination as the temperature drops. Another example is the expansion of a relativistic gas. We have seen how the velocity distribution of a gas is unaffected by changes in R. In the relativistic case there is, however, a change in the momentum distribution and hence a departure from thermal equilibrium. Entropy is thereby generated.

All these examples are only important for relatively short intervals in the thermal evolution of the universe.

The functional form of $R(t)$ must be determined from some dynamical gravitational theory such as general relativity. In the latter case there is no unique solution for R, but in practice only those models are taken seriously which expand from a finite time t in the past at which, formally at least, $R \to 0$. Whether or not such a singular state of infinite energy density should be taken seriously in general relativity is still a matter of some dispute, but most cosmologists are prepared to believe that the universe was at least in a state of extremely high compression about 10^{10} years ago. In this region the energy density of the radiation, which varies like R^{-4}, dominates over the rest energy of matter, which varies like R^{-3}, and it is easy to show that under these circumstances $R \propto t^{1/2}$ in general relativity. Moreover, the temperature rises extremely fast as we pass back towards the singular state. In fact, it can be shown (using (2)) that in the early stages the temperature is approximately equal to $10^{10}t^{-1/2}$ °K, where t is in seconds. Consequently, before about 10^{12} seconds the matter is highly ionised and strongly coupled to the radiation. Typical reaction rates all becomes very large compared with the expansion rate as $R \to 0$ and the whole system is therefore in thermal equilibrium at a very high temperature. Direct observational evidence of this thermal equilibrium state was found with the discovery that the whole universe is filled with black-body radiation to a high degree of uniformity, at a present temperature of 2.7 °K, which may be thought of as a relic of the initial fireball.

The reader may feel that this result is somewhat paradoxical. In laboratory thermodynamics, thermal equilibrium at uniform temperature and density is normally regarded as a maximum entropy condition. And yet, in the cosmological case, this structureless condition cannot be a state of maximum entropy otherwise further irreverisibility is impossible (apart from that due to the cosmological expansion, which we have already seen is rather unimportant) Evidently, the smooth primordial fluid wants to evolve into the highly structured condition of the world that we observe today. We are familiar with analogous situations in elementary phase transitions; for example, the condensation of a liquid from a gas, in which an entropy increase is accompanied by a growth of macroscopic structure. In cosmology, the "phase transition" is brought about by the force of gravitation, which plays a decisive role in the origin of irreversibility.

Heuristically, one may see that gravitation has a role to play in irreversibility by noticing that it is always attractive, unshielded and unsaturatable. Consequently, unlike the other forces of nature, it does not obey Lenz's law i.e. the effects it produces enhance rather than diminish its influence. Specifically, a mass will tend to accrete surrounding material, which in turn increases the mass and leads to greater accretion.

When gravitation is introduced into thermodynamics [6] a number of profound modifications appear connected with the fact that there is no true equilibrium configuration. Laboratory systems can always be maintained in equilibrium by manipulating the boundary conditions, but gravitating systems tend to collapse unless prevented by *internal* forces, such as the pressure gradients found inside stars, which produce a metastable relative equilibrium. But formally, the free energy of a gravitating system has no minimum, the final state of matter being not one of uniformity, but a black hole singularity of apparently infinite entropy. Even a system in metastable equilibrium does not tend to a uniform temperature throughout. Heat, being a form of energy, tends to flow to the bottom of potential wells (according to general relativity) so forming a temperature gradient.

There is a close connection between the tendency of gravitating systems to maximise their entropy by collapse and the onset of violent density fluctuations in a uniform fluid. It has long been the hope that ordinary $\sqrt{N}$ fluctuations might account for the existence of the observed hierarchy of structure in the world—star clusters, galaxies, galactic clusters etc., but this has not so far been achieved because these fluctuations are much too small to grow such large density perturbations in the time available. It appears necessary to postulate some sort of initial fluctuations of a non-thermal variety. Indeed, this is quite reasonable because the existence of a particle horizon in general relativity has the effect of isolating the universe into regions which are causally (and thermally) unconnected before a certain epoch, so there is some justification in expecting different regions to expand in an unrelated way initially. All that is required for our purposes is that the universe is smoother than a singularity. The fact that a finite universe exists at all is sufficient to reduce the entropy below the maximum (infinite) value of the singular state $R(t) \to 0$.[5])

The chain of processes consequent upon the formation and growth of inhomogeneities in the cosmological medium leading to the familiar forms of irreversibility that we encounter in physics is obvious in outline. The condensing masses undergo viscous drag, turbulence, friction and all the other usual forms of entropy generating processes. All of these tend to reduce the stability of the systems against further collapse. Most importantly, temperature gradients are set up, and heat flows occur. The organisation of matter into a highly structured state is compensated by entropy generation mainly in the form of electromagnetic radiation which flows away into space. Certain special regions (the centres of stars) then reach conditions

[5]) The reader should not confuse this remark with our earlier discussion in which the total entropy per comoving volume was shown to be roughly constant in the cosmological case. The present discussion is about *gravitational* thermodynamics, whereas the previous result was derived from ordinary laboratory thermodynamics and geometry, and did not include considerations of the gravitational field.

in which the low entropy atomic form of matter (hydrogen) becomes structured into heavier elements, again with the emission of compensatory high entropy photons (and neutrinos). Note that nucleosynthesis in the initial fireball generates very little entropy because it takes place under quasi-equilibrium conditions, but in stars there is an enormous temperature gradient between their radiating surfaces and the cold depths of space. This is maintained by the cosmological expansion which makes the universe act as a continuous sink of heat, and results in a great increase in entropy.

It is curious that although we usually think of the universe as a very cold place in this respect, there is also a sense in which it is very hot. Most of the entropy is in the form of primordial photons, and (as mentioned) there are 10^8 times more photons than baryons in the universe. The heat capacity of this radiation is enormously greater than that of the matter, and also about a hundred times greater than all the starlight. So in our discussion about enormous entropy generation in stars, the actual total increase in entropy represents a very small perturbation. Nevertheless, it is on the disequilibrium of this small perturbation that all life and structure depends.

Non-Entropic Irreversibility

There are many conspicuous examples of processes the irreversibility of which is non-entropic, though such processes may involve entropic irreversibility also. Examples of these processes are the retarded propagation of waves, the infinite dispersal of an exploding gas cloud into empty space and the destruction of the phase relationships between waves of different frequencies propagating in a dispersive medium.

Let us focus our attention on one of these. Suppose that we throw a stone into a large surface of water. We will observe a series of circular waves propagating radially from the point of disturbance (*retarded* disturbance). The reverse of this process is never observed to occur spontaneously, that is, we never encounter the accidental spontaneous concatenation of disturbances at the perimeter of a circle causing a convering coherent wave to shrink to a sharp point of extinction (*advanced* disturbance). Popper [7] has discussed this example, and pointed out that such a phenomenon would require organisation from a central point, because of the essential *coherent* nature of the reversed disturbance. Then the reverse of *this* centrally organised process could not occur. The spontaneity of the reverse process is essential, for we could indeed produce it easily ourselves by dropping a large circular object horizontally onto the water surface.

We could remove the spontaneity proviso by allowing the wave to propagate to infinity [1]. This will not occur with a water surface, but can happen for instance with light waves. The non-existence of *finite* spontaneous converging waves is easily explained thermodynamically. First we note that for a closed finite system in thermodynamic equilibrium both forms of irreversibility are absent. We may equally well discuss the retarded progress of a photon from atom A to atom B in terms of the advanced progress from B to A. As all correlations are absent in thermal equilibrium, diverging and converging wavefronts composed of large numbers of correlated photons do not occur. Next we note that when a diverging light wave or water wave reaches a boundary surface it becomes absorbed in the surface by various thermodynamic processes. The reverse situation (advanced radiation) would require co-operative enhancement of the wave amplitude through the surface, which is overwhelmingly improbable on thermodynamic grounds. However, for a disturbance which propagates to infinity this argument is no longer valid. It is true that a light wave will suffer some attenuation, but this is hardly relevant to the fact that we should require for the reverse process the apparently miraculous convergence of source free radiation from past infinity, coherently on a spherical surface. This inability of nature to supply correlated influences from widely separated regions of the universe in different directions has been stated as a "law of conditional independence" by Penrose and Percival [8].

Although called a law, this is really a boundary condition stating that are no converging coherent waves from the distant past. Why this should be so is still disputed. One obvious explanation is to simply deny the existence of *any* source free radiation, at either past or future infinity. This is the position adopted in the Wheeler–Feynman theory to be discussed below (as we have seen, advanced radiation from *sources* in the past converging onto accelerated charges is precluded on thermodynamic grounds). In any case, the whole concept of source free radiation conditions in the past is suspect near the singular region $R \to 0$. We cannot decouple the radiation from the matter as the density rises indefinitely. These and similar problems have been discussed by Sciama and Ellis [9].

Sciama [10] has also pointed out that the general existence of advanced electromagnetic radiation leads to an energy density divergence in some cosmologies. For instance, consider an everexpanding Einstein–de Sitter world with $R \propto t^{2/3}$, filled with stars of luminosity $L(t)$. Advanced radiation from these objects would be blue-shifted at earlier times. If the rate of energy emission per unit volume is proportional to $L\varrho$ where ϱ is the average density of stars, and the rate of change of energy density due to the cosmological expansion is $4\dot{R}/R$, the net total energy density at time t due to all the advanced radiation from all the sources in the future is clearly proportional to $R^{-4}(t) \int_t^\infty L(\tau)\, \varrho(\tau)\, R^4(\tau)\, \mathrm{d}\tau$. As ϱR^3 is constant this expression is proportional to $R^{-4}(t) \int_t^\infty L(\tau)\, R(\tau)\, \mathrm{d}\tau \propto t^{-8/3} \int_t^\infty L\tau^{2/3}\, \mathrm{d}t$, which diverges if $L(t)$ falls slower than $t^{5/3}$ as $t \to \infty$. Therefore, in cosmologies which have galaxies cooling slower than this rate in the distant future, advanced radiation would produce an unacceptable divergent energy density.

Actually, an explanation of the non-existence of fully advanced electromagnetic radiation is not really appropriate to a discussion on irreversibility anyway, for such fields are still time asymmetric. What is more important is why a symmetric mixture of advanced and retarded solutions are not observed. The main difficulty with the whole discussion of a mixture of advanced and retarded solutions is that of self-consistency. The concept of an initial value problem is inapplicable in the case of a mechanical system for which the present behaviour depends on complicated interrelations between past and future. In general we cannot say what the behaviour of an arbitrary system of charges interacting symmetrically through advanced and retarded fields might be, or whether the solution is unique anyway. Curiously enough, however, some self-consistent solutions can be found, and are described in a paper by Wheeler and Feynman [11]. All such solutions require the existence somewhere of a complete absorber for the radiation, either on the future or past light cone. One such solution will give only the full retarded radiation fields of experience.

To understand how this time asymmetric field can arise from an interference between time symmetric fields, consider a charged particle i located inside an opaque box, and accelerated for a brief instant. The observed retarded field of the particle will set into motion the charges in the walls of the box. These charges will have associated retarded and advanced fields, the latter of which will react back on the original charge i simultaneously with its acceleration. This charge therefore experiences a force. Moreover, the same field will interfere with the retarded and advanced fields of i. It is easy to show by a direct calculation that provided the walls of the box are truly opaque the above mentioned force is just the usual radiative damping force, while the interference of the advanced response field from the walls just cancels the advanced field from i and reinforces its retarded field to the full strength assumed originally. The argument is therefore self-consistent.

The temporal asymmetry arises in this theory from the nature of the opacity of the absorbing material. The damping processes which produce this opacity occur with an asymmetry already fixed from thermodynamics, so that it is only the retarded radiation from i that can be absorbed and generate a response. The advanced fields cannot provide such a self-consistent solution.

Mathematically, we can obtain the above results in a general way. The inhomogeneous Maxwell equation may be solved to obtain a particular solution, to which may be added any further solution of the homogeneous equation, to satisfy the boundary conditions. Specifically, we may add to the time-symmetric solution $\frac{1}{2}F_{\text{retarded}} + \frac{1}{2}F_{\text{advanced}}$ the homogeneous solution $\frac{1}{2}F_{\text{retarded}} - \frac{1}{2}F_{\text{advanced}}$ to obtain F_{retarded}. Now we know that there are no (total) fields outside the opaque box due to disturbances occurring within, so that the above homogeneous solution also vanishes there. But it is well known that the only homogeneous solution which vanishes permanently on a surface is the trivial one i.e. the field which is everywhere zero. Consequently, there can be no difference in using the time-symmetric fields or F_{retarded}, as they differ at all points only by a zero quantity. It is important to realise that in arriving at this result, F is the *total* field of all the charges set into motion by the original disturbance, including those due to the particles of the box.

Because the advanced fields cancel by interference, and the retarded waves are absorbed by the box, there can be no real photons in this theory i.e., no "free" radiation at infinity. Consequently the field degrees of freedom may be entirely removed from the mechanical description of the system. The quantisation of such an action-at-a-distance theory then requires special techniques [12].

The question of whether our universe does behave like the inside of an opaque cavity depends on the behaviour of matter and radiation as $t \to \infty$. This brings us to a discussion on the end of the world.

The End of the World

In a closed system with constant constraints, it is well known that on reaching a state of maximum entropy, further comparable change is impossible on time scales less than the Poincaré recurrence time. An application of this to a static universe predicts that eventually all physical systems will "run down" to an equilibrium situation, and all known processes will be balanced by their inverses. The flow of time will have no meaning and the world will remain unchanged for countless aeons. This state of affairs is popularly known as the heat death of the universe, and is felt by many to be philosophically unpalatable. Many attempts have been made to avoid the heat death by one means or another. Some of these attempts will now be described.

First we note that continued expansion of the universe will not really affect matters. Only inconspicuous entropy generation will continue at a diminishing rate, such as the slow flow of heat from radiation to matter already discussed.

We can calculate [13] the cooling rate of ions coupled to the radiation field in a model which expands like t^n. For $n \geqq 1$ it turns out that the coupling is negligable and the ion temperature T_i drops like the usual R^{-2} for uncoupled matter. For $n < 25$ the coupling is so strong that the ions cool like the radiation field: $T_i \sim R^{-1}$. For intermediate models we have cooling at a rate $T_i \sim R^{-p}$ where p and n are related by $\frac{3}{2}p = 4 - \frac{1}{n}$. These cooling rates are shown in Fig. 4 where we have plotted p against n. Actually, it is found that (for models which expand slower than $t^{1/2}$) there are no unrecombined ions left as $t \to \infty$. The ionic recombination will also generate entropy.

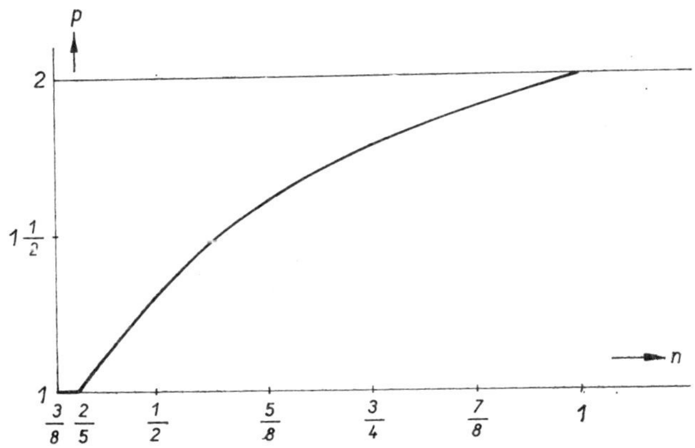

Fig. 4. Ionic cooling rates $T_i \sim R^{-p}$ where $R \sim t^n$.

Most of the conspicuous forms of irreversibility are however largely unaffected by the expansion, and will die out in spite of it. Radiation emission by stars will continue for a while, the energy density of starlight becoming comparable with that of the black body background after about $\frac{1}{n} \times 10^{\frac{12+10n}{1+n}}$ years. Systems that depend on gravitational contraction and nucleosynthesis for their entropy production will eventually use up all available energy and may collapse to the ultimate high entropy condition of a black hole.

The fate of electromagnetic radiation in the heat death is easy to determine. If the cross section for photon absorption is independent of frequency, as in the case of macroscopic objects such as stars, the probability of a photon being absorbed in a time dt is proportional to $\varrho\, dt$ (actually $1 - e^{-\varrho dt}$) where ϱ is the density of absorbing surfaces. Consequently absorption will eventually be complete if $\int \varrho\, dt$ diverges i.e., if ϱ falls slower than t^{-1}. In matter conserving cosmologies ϱR^3 is constant, so that the limiting case of an opaque model is the Dirac cosmology, $R \propto t^{1/3}$. All the ever expanding general relativity models expand faster than this and are hence transparent. (This remains true even if we allow for absorption by free ions via inverse bremmstrahlung. The limiting model is then $R \propto t^{2/5}$.) Only the Dirac model therefore provides the correct conditions for recovery of retarded radiation from the time symmetric solutions, using the Wheeler – Feynman theory.

In an attempt to prevent an ever expanding universe from undergoing the heat death Hoyle, Gold and Bondi [14] postulated the continual creation of matter. Usually this is considered to be spread uniformally throughout the universe and is hence in a low entropy state; capable of gravitational condensation and irreversibility. By injecting negative entropy continually into the world from "without" it is possible to maintain the world in a steady state, with constant entropy density throughout. The heat death is thereby avoided and the universe behaves like a gigantic perpetuum mobile. The steady state model as such is generally thought to have been discredited by observations, though the possibility of using continual creation even in evolutionary models may still appeal to some [15]. Alternatively, one could conjecture a multiply connected topology, in which high entropy collapsing matter falling into a black hole erupts elsewhere as a low entropy "white hole". There is some evidence for violent ejection of matter from galactic nuclei.

Another possibility of avoiding the heat death is by killing the universe in a different way. Such a situation occurs in the closed Friedman models which expand for a time to a maximum value of R, after which a recontraction sets in, eventually compressing the matter back into a dense (or singular) condition after a finite time.[6]) This cataclysmic expiry of the world may be more spectacular, but is hardly less philosophically disturbing than the heat death. Consequently, some writers have suggested that each cycle of expansion and contraction is followed by another, so sharing with the steady state model the advantage of avoiding a beginning and an end to the world, but without the necessity for the continual creation of matter. The trouble is that although in laboratory thermodynamics the entropy of a closed system must eventually reach a maximum after which no change can occur, in cosmology the changing gravitational constraints accompanying the expansion and recontraction allow the entropy to increase without limit from cycle to cycle [4]. Any observer should therefore experience an infinite entropy.

A way of avoiding this problem has been suggested by Gold [16], who considers the role of the cosmological expansion in the origin of irreversibility to be crucial. Gold suggests that

[6]) This type of model also satisfies the Wheeler—Feynman condition of opaqueness.

should we live in an oscillating cosmology of the type considered above, then the cycle of expansion and contraction might be reversible. Earlier we saw how a space uniformally filled with radiation could behave reversibly. The outward flow of photons from a small region which occurs on expansion is simply reversed on contraction. No entropy is generated if the expansion is uniform because there are no temperature gradients. This reversible flow of photons is in sharp distinction to the flow of radiation away from the surfaces of stars. In this case there are temperature gradients around local inhomogeneities, and the flow of radiation does generate entropy, so cannot be reversible on thermodynamic grounds. Moreover, this flow of radiation into empty space has no direct connection with the cosmological expansion. Given that space is at a relatively low temperature (e.g. 3 °K), then this outward flow is bound to occur when gravitational collapse, nucleosynthesis etc. set up local regions of higher temperature.

Nevertheless, Gold maintains that even this latter type of radiation flow would be reversed at the changeover from expansion to contraction. Because these processes are manifestly entropy generating, we have to make the *additional* postulate that the direction of entropy change is also reversed relative to the expanding phase, though (as we just mentioned) there is no particular reason to associate the two processes (cosmological expansion and entropy generation). As this changeover would have to occur long before the universe reaches thermal equilibrium this is a somewhat startling idea.

A very real objection to this hypothesis also comes from a consideration of non-entropic irreversibility. A reversal of the time direction of entropy change cannot affect non-entropic processes, and in particular the outward, diverging spherical light waves remain diverging when they cross the "turning point" of maximum expansion. But the reversed energy generation in the contracting phase requires *converging* coherent radiation onto stars etc. We must be careful not to confuse this process with the rise in sky temperature as the contraction proceeds [17]. This will occur for two reasons. Firstly, the black body background temperature grows like R^{-1} as R decreases. Secondly, the starlight emitted by *approaching* galaxies becomes blue shifted and hot. Eventually (but only near the final collapsed state, $R \to 0$) the sky will be hotter than the stars. We should therefore have a flow of radiation *from* space into the stars (which would soon vapourise). But this process is equally entropy generating, and would only occur if we kept our usual thermodynamic sense of entropy change in the contracting phase also. Moreover, the flow of radiation from space into the stars does not occur in converging waves onto special inhomogeneities, but rather by increasing the radiation uniformally, throughout all space. Note that a given observer will not be aware of the contraction of the universe by observation of distant galaxies until well into the contracting phase, because of the lag in light reaching the observer from these regions; and because the black body background temperature is so low anyway the sky will still be very dark and cold compared with the stars for a long time after the maximum expansion. Thus in no sense is the expansion–contraction cycle reversible.

We could attempt to remedy the situation by postulating that the diverging wavefronts from one half of the cycle conspire to form converging wavefronts in the other half, but it is hard to do this in detail because most of the radiation available is the blackbody background, which is too low a temperature to form into starlight. Another way of putting this to say that most photons from stars that travel into the other phase will miss the stars there and plunge into the fireball near $R \to 0$. By this time they will be much more energetic (due to the blue shift $\omega \sim R^{-1}$) than the blackbody radiation, and cannot therefore be viewed as part of it. We also have an unstable situation for radiation. Normally, electromagnetic waves are damped away on encountering matter, but in this model universe, most of the radiation crosses the boundary and travels into the opposite phase where, because entropy change is reversed, it is not damped, but enhanced. There is therefore an inbuilt instability in such a model.

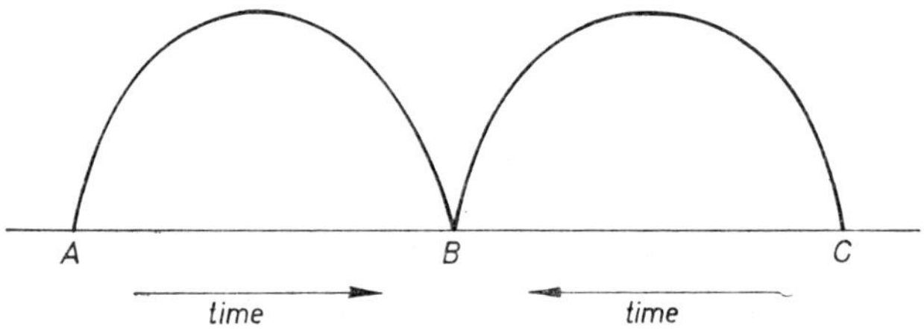

Fig. 5. A two cycle time-symmetric universe.

In spite of all these objections to Gold's suggestion, the idea of a reversible universe is still philosophically attractive. There is another way [18] in which the same result could be achieved, without the difficulty of joining on two opposing phases in a non-equilibrium situation. To see this consider Fig. 5, which shows *two* successive cycles starting and ending in very dense states *A*, *B* and *C*. As have seen, the region *B* separating the two cycles will be in thermal equilibrium, and so will be reversible. Consequently there is no reason to suppose that the direction of irreversible processes will be the same on either side of *B*, and we might expect a situation where the entropy increases to the left of *B* and decreases to the right. There is now no problem about nonentropic irreversibility, because all coherent wavefronts are destroyed in the thermalisation process at *B*. That is, the starlight from *AB* appears as a thermal background in *BC* and vice versa. This statement is subject to empirical verification. Making some simple assumptions about the rate of starlight production, and using present observational data, we can compute [18] that at our epoch, the thermalised starlight from the "next" cycle would have a temperature of about 3 °K, in agreement with observation. This number has hitherto remained unexplained.

We should suppose that an observer in either cycle would interpret the world identically. Although relative to each other their time asymmetry is reversed, the human brain presumably participates in the thermodynamic behaviour of the world, and an observer could not distinguish observationally between the cycles on thermodynamic or electrodynamic grounds. This may not be true of the laws governing K-mesons. It may be necessary for an observer to interchange his definition of matter and antimatter before these laws are also identical.

As a final refinement, we could replace the unending series of entropy reversing cycles with a closed time axis, where the states *A* and *C* are identified. The world then has the topology of a circle, and, whilst having a finite lifetime, has no beginning and no end.

Summary

It is a curious fact that such a fundamental aspect of our experience as the asymmetry of time should appear to depend on small perturbations to the world. The gravitational instability of density fluctuations leads eventually to all the major sources of entropy generation, and all the world structure that we observe. The existence of retarded radiation is less easy to account for, but the time symmetric electrodynamics of Wheeler and Feynman satisfactorily attributes this to thermodynamic irreversibility. However this theory demands that the universe be opaque, a condition only satisfied by the Dirac and oscillating closed Friedmann models. The ultimate fate of matter appears to be a type of heat death by collapse to black holes, while the fate of radiation depends on the cosmology. An alternative – the fascinating possibility of a "regenerating" cyclic world with a closed time axis – could be provided by a closed Friedmann model which bounces out of the final collapse into a "time reversed" phase.

The ultimate origin of most irreversibility is the low entropy condition of the smooth universe at its origin a finite time ago.

References

[1] A. Grünbaum: Philosophical Problems in Space and Time. Routledge & Kegan Paul, London 1964.

[2] H. Reichenbach: The Direction of Time. University of California Press, Berkeley 1956.

[3] E. Schrödinger: Irreversibility. Proc. Roy. Irish Acad. *53* (1950), 189—195.

[4] R. C. Tolman: Relativity, Thermodynamics and Cosmology. Clarendon Press, Oxford 1944; chapter 10.

[5] D. Layzer: Cosmic Evolution and Thermodynamic Irreversibility. IUPAC International Conference on Thermodynamics—Cardiff (P. T. Landsberg, Ed.). Butterworths, London 1970.

[6] W. C. Saslaw: Gravithermodynamics I. M. N. R. A. S. *141* (1968), 1—25.
W. C. Saslaw: Gravithermodynamics II. M. N. R. A. S. *143* (1969), 437—459.

[7] K. Popper: The Arrow of Time. Nature *177* (1956), 458—458.
K. Popper: Irreversibility and Mechanics. Nature *178* (1956), p. 382.
K. Popper: Irreversible Processes in Physical Theory. Nature *179* (1957), p. 1297.
K. Popper: Irreversible Processes in Physical Theory. Nature *181* (1958), 402—403.

[8] O. Penrose and I. C. Percival: The Direction of Time. Proc. Phys. Soc. *79* (1962), 605—616.

[9] G. Ellis and D. W. Sciama: Global and Non-global Problems in Cosmology. In: General Relativity-Papers in Honour of J. L. Synge (L. O'raifeartaigh, Ed.). Clarendon Press, Oxford 1972.

[10] D. W. Sciama: Retarded Potentials and the Expansion of the Universe. Proc. Roy. Soc. Ser. A. *273* (1963), 484—495.

[11] J. A. Wheeler and R. P. Feynman: Absorber Theory of Radiation. Rev. Mod. Phys. *17* (1945), 157—181.

[12] P. C. W. Davies: Extension of Wheeler—Feynman Quantum Theory to the Relativistic Domain: I Scattering Processes... II Emission Processes. J. Phys. A. *4* (1971), 836—845; *5* (1972), 1025—1036.

[13] P. C. W. Davies: Is The Universe Transparent or Opaque? J. Phys. A. *5* (1972), 1722—1737.

[14] F. Hoyle: A New Model For The Expanding Universe. M.N.R.A.S, *108* (1948), 372—382.
H. Bondi and T. Gold: The Steady-State Theory of the Expanding Universe. M.N.R.A.S. *108* (1948), 252—270.

[15] F. Hoyle and J. V. Narlikar: Cosmological Models In a Conformally Invariant Gravitational Theory I;... II. M.N.R.A.S. *155* (1972), 305—321; 323—335.

[16] T. Gold: The Arrow of Time. Amer. J. Phys. *30* (1962), 403—410.
[17] M. J. Rees: The Collapse of the Universe: An Eschatological Study. The Observatory *89* (1969), 193—198.
[18] P. C. W. Davies: Closed Time as an Explanation of the Black Body Background Radiation. Nature *240* (1972), 3—4.

CAN INFORMATION THEORY RECEIVE A RELATIVISTIC EXPRESSION?

O. Costa de Beauregard

Bergson recounts somewhere that, having once been asked what to-morrow's novel would look like, he answered that he did not know, for, had he known, this would imply that he had written the novel.

Well, up to now, only this title of the novel exists. Notwithstanding Bergson's serious warning let us try to outline what it should like.

Speaking of *relativity theory* implies at once that the time concept is taken in a strongly physical sense, and not in some remotely abstract fashion as may be the case in an academic version of probability theory or information theory. We will be dealing with some manifestly covariant form of statistical mechanics.

It so happens that Jaynes [1] partly inspired, as it seems [2], by Cox [3], produced a self-contained general formalism of statistical mechanics, yielding an extremely concise and transparent deduction of all the essentials. Later Tribus [4] published elegant presentations of this doctrine and Katz [5] has developed it in the form of a treatise.

Jaynes' prescription for obtaining the significant probability distribution is derived straight from information theory and reads: Maximize the entropy

$$S \equiv -\Sigma\, p_i \ln p_i \tag{1}$$

subject to

$$\Sigma\, p_i = 1 \tag{2}$$

and whatever contraints result from the physical situation. For instance, in classical statistical mechanics, the mean value $\langle E \rangle$ of the energy E, and the mean values $\langle n_m \rangle$ of the numbers of different chemical molecules m, are given, whence

$$\Sigma\, p_i E_i = \langle E_i \rangle, \tag{3}$$

$$\Sigma\, p_i n_{mi} = \langle n_m \rangle \tag{4}$$

etc... Then the well known technique of Langrange multipliers yields, through the definition of the partition function

$$\pi = \ln \sum_i e^{-\beta E_i - \sum_m \beta_m n_{mi}}, \tag{5}$$

the following expressions for the entropy, the mean energy, the mean concentrations, etc...

$$S = \pi + \beta \langle E \rangle + \sum_m \beta_m \langle n_m \rangle, \tag{6}$$

$$\langle E \rangle = -\partial\pi/\partial\beta, \tag{7}$$

$$\langle n_m \rangle = -\partial\pi/\partial\beta_m. \tag{8}$$

Of course, the prescription and the formulas are nothing more than what is found in every classical exposition of equilibrium statistical mechanics. What is new is that the line of reasoning is reduced to its bare skeleton, plus the motive muscles enabling it to be applied directly for solving any problem within its potential field. The huge obesity of classical statistical mechanics – that is, the extensive considerations on ensembles and so on – is completely gone, much obscurity in applications or extentions has also disappeared, and much clarity has been gained.

As for relativistic covariance there is *a priori* "no problem". The exponent in formula (5) must be a scalar. Therefore, if we insert instead of the energy E the energy-momentum 4-vector E^λ, with $\lambda = 1, 2, 3, 4$, the "inverse temperature" β inside the scalar product $\beta_\lambda E^\lambda$ will show up as a 4-vector β_λ. If we need to introduce the six-component angular momentum $E^{[\lambda\mu]}$, a corresponding skew-symmetric inverse temperature $\beta_{[\lambda\mu]}$ will come in. As for the β_m's, they will of course remain scalars, etc...

Therefore, the relativistic covariance should not bring in *per se* any serious problem.

In a physically significant information theory, we also need a covariant formalism for information transfer. At this end wave propagation is ideally suited. Moreover, as we should not be restricted to electromagnetic waves, it is only natural to turn towards wave mechanics, that is, quantum mechanics.

So the question naturally arises: can we find, in the quantum mechanical formalism, an acceptable definition of entropy "corresponding" to the classical one? The answer is known to be *yes* – through J. von Neumann's definition [6] of the "density matrix" P and his "ensemble" presentation of quantum mechanics. The entropy is then defined as

$$S = -\text{Trace}\,(P \ln P). \tag{9}$$

It so happens that in an important, but rarely quoted article of 1937, Elsasser [7] inspired by Fisher [8] has applied, inside the von Neumann scheme, the very prescription later used by Jaynes, in order to produce a consistent theory of the quantum mechanical measuring process. In Elsasser's words [7, p. 992] this was an information theory approach. He also states that, in his line of thought, the density matrix should be used for describing not only ensembles, but even individual systems.

So one more of the things we are needing is already there: an anticipated sketch of the use of Jaynes' formalism inside the scheme of J. von Neumann's quantum mechanical density matrices *plus* an explicit application of it by Jaynes himself.

The next thing we need is a relativistically covariant formalism for the scalar product, the normalization and orthogonality formulas, the reciprocal Fourier transforms, all of which are the everyday bread of quantum mechanics, and thus the basis for constructing, in any specific case, the von Neumann density matrix.

Of course, since the celebrated works of Tomonaga [9], Schwinger [10], Feynman [11] and Dyson [12], everybody knows that a beautiful formalism exists for relativistic quantum electrodynamics, and some other specifications of quantum field theory. What we need for our purpose is something more down to earth: it is a relativistically covariant presentation of the "first" rather than of the "second quantized" mechanics.

It so happens that this very problem has been with me for many years, and that, after writing a few articles on it [13], I have expanded the whole subject in a booklet [14]. Wightman and Schweber [15] have also considered the problem.

So what seems to be the last necessary piece of our puzzle is also there: an explicitly covariant formulation of "first quantization", including a theory of Green's functions or "propagators", and a resolution of the Cauchy problem in the form of the expansion of a wave function on a complete set of orthogonal propagators.

So then what remains to be done? At this point one should remember Bergson's warning: building up an operational physical theory is not rattling on or waving hands. That is, building up a theory is *never* merely putting together pieces of a puzzle; one can never avoid using sandpaper – to say the least. Therefore it is *only* when building up the whole scheme that one will discover what it truly looks like – and almost certainly one will discover unexpected implications.

But man's mind is thus made that he cannot help trying to guess what unexpected things he should expect. In my opinion, there is one information transfer problem in space-time that is a real big one: the one pertaining to the so-called Einstein-Podolsky-Rosen [16] paradox, also considered by Schrödinger [17] and by Furry [18]. As I have already expressed [19] my feelings in this respect I will not do it again here. I will simply draw attention to a very interesting recent paper by Moldauer [20] entitled "Is there a quantum mechanical measurement problem?", the implication being that there is none. The case is made very strongly, and formulas for conditional probability are given. Therefore the paper is important to us.

However, notwithstanding Moldauer's dazzling exorcism of ghosts, I believe that the problem of information gain (knowledge), of information use (action), and of information transfer by means of waves still contains, in quantum mechanics, an unsolved mystery.

The point is clearly made by Hooker [21] and, in my opinion, brings us straight back to London and Bauer's [22] philosophy. In the Einstein-Podolsky-Rosen "paradox" the problem is neither prediction, nor even retrodiction, but truly *telediction*, along a spacelike separation built up from two timelike separations making a sort of Feynman zigzag [19]. It is *not* relevant for observer A to think of the other distant observer B as going to perform, or having already performed, his measurement when A is performing his own one, because this order in time is a *relative* and not an *absolute* one.

The emphasis here is on time symmetry of elementary processes. This calls for full consideration of the *intrinsic* symmetries between statistical prediction and retrodiction, retarded and advanced waves, information as knowledge and as volition, as discussed by quite a few authors [23].

To conclude, I do hope that a student of mine will show up and be ready to tackle these intertwined problems, because I certainly would like to look over his shoulder while the work is going on.

Also, I should like to say that the names I have cited in my paper are certainly not the only ones that could have been quoted. The approach to probability problems and to statistical mechanics, either classical or quantal, has truly been, either explicitly or implicitly, "in the air" for all these past years.

References

[1] E. T. Jaynes: Information Theory and Statistical Mechanics (I and II). Phys. Rev. *106* (1957), 4, 620—630 and *108* (1957), 2, 171—190.
E. T. Jaynes: Information Theory and Statistical Mechanics. In: Brandeis Lecture, Benjamin Inc., New York 1962, 181—218.

[2] E. T. Jaynes: The Algebra of Probable Inference. Amer. Journ. Phys. *31* (1963), 66—67.

[3] R. T. Cox: Probability, Frequency and Reasonable Expectation. Amer. Journ. Phys. *14* (1946), 1, 1—13.
R. T. Cox: The Algebra of probable Inference. The John Hopkins Press, Baltimore 1961.

[4] M. Tribus: Information Theory as the Basis for Thermostatics and Thermodynamics. Journ. Appl. Mech. (March 1961), 1—8.
M. Tribus: Information Theory and Thermodynamics. In: Boelter Anniversary Volume. McGraw Hill Book Co., New York 1963, 348—368.
M. Tribus and R. B. Evans: A minimal Statistical Mechanics from which Classical Thermodynamics may be derived. In: A Critical Review of Thermodynamics (E. B. Stuart, B. Gal-Or, A. J. Brainard, Eds.). Mono Book Corp., Baltimore 1970, 429—438.
M. Tribus: Rational Descriptions, Decisions and Designs. Pergamon Press, New York 1969.

[5] E. Katz: Principles of Statistical Mechanics. W. H. Freeman & Co., San Francisco 1967.

[6] J. von Neuman: Mathematical Foundations of Quantum Mechanics. Princeton Univ. Press, Princeton 1955.

[7] W. M. Elsasser: On Quantum Measurements and the Role of the Uncertainty Relations in Statistical Mechanics. Phys. Rev. *52* (1937), 987—999.

[8] R. A. Fisher: Inverse Probability. Proc. Camb. Phil. Soc. *26* (1929), 588—535.
R. A. Fisher: Inverse Probability and the use of Likelihood. Proc. Camb. Phil. Soc. *28* (1932), 3, 257—261.

[9] S. I. Tomanaga: On a relativistic invariant formulation of the Quantum Theory of Wave Fields. Prog. Theor. Phys. *1* (1946), 2, 27—42.

[10] J. Schwinger: Quantum Electrodynamics I. A Covariant Formulation. Phys. Rev. *74* (1968), 10, 1939—1961.

[11] R. P. Feynman: The Theory of Positrons. Phys. Rev. *76* (1949), 6, 749—759.
R. P. Feynman: Space-time approach to Quantum Electrodynamics. Phys. Rev. *76* (1949), 6, 769—789.

[12] F. J. Dyson: The radiation Theories of Tomonaga, Schwinger and Feynman. Phys. Rev. *75* (1949), 3, 486—502.

[13] O. Costa de Beauregard: Sur la Dynamique Relativiste du Formalisme quantique en Théorie de Dirac. Comptes Rendus *225* (1947), 5, 626—629.
O. Costa de Beauregard: Développement d'une Théorie de Marcel Riesz. Forme covariante de la fonction de distribution de l'énergie de l'électron libre. Comptes Rendus *232* (1951), 9, 804—806.
O. Costa de Beauregard: Covariance relativiste à la base de la Mécanique Quantique. Journ. Phys. Rad. *15* (1954), 12, 810—816.
O. Costa de Beauregard: Covariance relativiste à la base de la Mécanique Quantique. Journ. Phys. Rad. *16* (1955), 10, 770—780.

[14] O. Costa de Beauregard: Précis de Mécanique Quantique Relativiste. Dunod, Paris 1967.

[15] A. S. Wightmann and S. S. Schweber: Configuration Space Methods in Relativistic Quantum Field Theory Phys. Rev. *98* (1955), 812—837.
[16] A. Einstein, B. Podolsky and N. Rosen: Can Quantum Mechanical Description of Physical Reality be Considered Complete? Phys. Rev. *47* (1935), 777—780.
[17] E. Schroedinger: Die gegenwartige Situation in der Quantenmechanik I, II, III. Naturwiss. *23* (1935), 807—812, 823—828, 844—849.
[18] W. M. Furry: Note on the Quantum-Mechanical Theory of Measurement. Phys. Rev. *49* (1936), 393—399.
[19] O. Costa de Beauregard: Deux Problèmes en Epistémologie du Temps de la Physique. Revue Intern. Philos. *61—62* (1962), 3/4, 1—19.
O. Costa de Beauregard: Le Paradoxe de Corrélations d'Einstein et de Schrödinger et l'éqaisseux temporelle de la Transition Quantique. Dialectica *19* (1965), 3—4, 280—289.
O. Costa de Beauregard: Temporal Asymmetry in Thermodynamics and Cosmology. In.: Proc. Intern. Conf. on Thermodynamics (P. T. Landsberg, Ed.). Butterworths, London 1970, 539—548.
[20] P. A. Moldauer: Is there a Quantum Measurement Problem? Phys. Rev. *D5* (1972) 4, 1028—1038.
P. A. Moldauer: A reinterpretation of von Neumann's Theory of Measurement. Found. of Physics *2* (1972), 1, 41—47.
[21] C. A. Hooker: Against Krip's resolution of two paradoxes in Quantum Mechanics. Philosophy of Science *38* (1971), 3, 418—428.
[22] F. London et E. Bauer: La Théorie de l'Observation en Mécanique Quantique. Herman, Paris 1939.
[23] O. Costa de Beauregard: No paradoxe in the Theory of Time Anisotropy. Studium Generale *24* (1971), 10—18 (with numerous bibliographical references).

ENTROPY AND RELATIVITY THEORY

Richard Schlegel

Considering the profound effect which the theory of relativity has had upon our ideas about the physical world, and, also, taking into account the power and utility of entropy calculations in physical theory, it is natural to ask, "In what ways does relativity theory modify our concept of entropy?" The answer to the question, we shall see, is that at present no pronounced change has been found to be necessary; there are not even the transformation-effects of relative motion which special relativity theory gives for such dynamical variables as space, time, and mass-energy. The relationship between the entropy-increase principle and the forward direction of time does, however, open up certain new possibilities for the interpretation of relativity theory; these relate particularly to the well-known relativistic time-dilation, and will be discussed with respect to changes of clock rates with motion. Empirical studies of these changes are seen as having new implications for decisions about the constitutive role of time in physical processes.

I. Relativistic Thermodynamics

In the early years of relativity theory the problem arose of how thermodynamic quantities such as temperature, pressure, heat, work, or entropy should be transformed under the conditions of relative motion that require a Lorentz transformation on space and time variables. In response, Planck, von Laue, Einstein, and Tolman developed a relativistic thermodynamics;[1]) two notable results of their work are the transformation properties of heat and temperature:

$$dQ = dQ_0(1 - v^2/c^2)^{1/2}, \tag{1}$$

$$T = T_0(1 - v^2/c^2)^{1/2}. \tag{2}$$

In Equations (1) and (2), T_0 is an equilibrium rest temperature in a physical system moving

[1]) For an exposition of this "orthodox" treatment see either M. von Laue [1] or R. C. Tolman [2].

at speed v relative to an observer for whom the temperature becomes T; c is the speed of light. Likewise, $\mathrm{d}Q$ and $\mathrm{d}Q_0$ are differential heat elements for moving and rest observers respectively.

It is apparent that for entropy differential $\mathrm{d}S = \mathrm{d}Q/T$, the preceding equations lead to the invariance relationship,

$$\mathrm{d}S = \mathrm{d}S_0. \tag{3}$$

That is, the effects of relative motion as expressed by the Lorentz transformation leave the entropy of a system unchanged.

However, the relations (1) and (2) have not gone unchallenged, and in the past decade or so there has been a lively discussion in the physics journals about what are the correct equations of relativistic thermodynamics. In particular, work by Ott [3] led to a cogent alternative system in which the analogues of (1) and (2) are:

$$\mathrm{d}Q = \mathrm{d}Q_0/(1 - v^2/c^2)^{1/2}, \tag{4}$$

$$T = T_0/(1 - v^2/c^2)^{1/2}. \tag{5}$$

Ott's work has been widely favored, but many other possible formulations have also been presented. An analysis of what is involved in the various definitional schemes of relativistic thermodynamics, as well as an historical survey, have been given by L. A. Schmid [4].

II. Entropy Invariance

For our purposes there is the significant result that in all of the systems of relativistic thermodynamics which have been presented the total entropy of a system is found to be a scalar–invariant, just as in the original Planck-Einstein-von Laue-Tolman system. Tolman argues that such an invariance is intuitively reasonable with either the heat or statistical definition of entropy [2, pp. 157–158]. Thus, given two systems at rest, each with entropy S, it should be possible to put one of them into a state of relative motion adiabatically, that is, with no transfer of heat to the system and hence $\mathrm{d}Q = 0$. But with $\mathrm{d}Q = 0$ we of course also have $\mathrm{d}S = \mathrm{d}Q/T = 0$. Or, using the statistical definition, $S = k \log W$, where k is Boltzmann's constant and W is the number of microscopic arrangements of elements of the system that are possible for a given thermodynamic state, we can argue that the relative probability of a given state (that is, the number of corresponding microstates) should not change in consequence of motion between the observer and the system.

III. Relativistic Time Change

The association[2]) of the passage of time with entropy increase is an objective relationship which may be utilized as a mark of the direction of time, aside from the subjective sense of

[2]) See the papers of R. Schlegel [5], L. Brillouin [6], I. Szumilewicz [7], and J. Zeman [8].

past-present-future which serves as the basic ground of our time sense. An apparent inconsistency arises, however, when we bring together the time-entropy relationship, the relativistic time transformation, and the relativistic invariance of entropy. In broad terms, the problem is that we can define an entropy function relating to the time of a clock, and, further, we have the presumption that entropy is independent of relative motion; yet, the special theory of relativity requires (and there is an abundant body of empirical evidence) that time rates change in consequence of motion relative to an observer.

In order to pose the problem in greater detail we assume that we have a macroscopic clock for which the entropy may be determined. The clock, for example, might operate by the slow passage of a gas, in a state of quasi-equilibrium, between two chambers, one initially empty and the other containing the gas at high pressure. We calibrate the entropy of the gas clock against a standard clock, so that we have $t = f(S)$, where t is the time as given by the entropy clock. Now, the function $f(S)$ must be the same in any coordinate system, independent of relative motion, since otherwise we would violate the Principle of Relativity (and could determine a state of uniform motion by measurements that are altogether internal to a system).

Further, suppose we have a set of the entropy clocks, all of similar construction, and at rest at various points along the x-axis. Yet another similar clock is set into motion with velocity $v = v_x$, at $x = 0$ and at the instant when all the clocks read $t = 0$. Now, for every increment δt each clock will gain an entropy increment δS, with $\delta t = f(S + \delta S) - f(S)$. However, because of the Lorentz invariance of entropy, the entropy value S of a clock at rest at a given x point must be the same as the entropy S of the moving clock as it passes that point. Otherwise we would have a $\mathrm{d}S/\mathrm{d}v \neq 0$, which is contrary to the assumed $S = S_0$. We conclude, then, that since $t = f(S)$ in all coordinate systems we must have an invariant clock time t for our entropy clocks. That is, relative motion does not change the rate of the entropy clock, and the moving clock will give the same reading at any x point as does the rest entropy clock at that point.

In contrast, the special relativity theory requires the time transformation, $t' = (t - vx/c^2)/(1 - v^2/c^2)^{1/2}$, between time t' of the moving clock and time t of the rest clock at x. Is this Lorentz transformation not applicable to the moving macroscopic entropy clock? As already noted, the relativistic time transformation has been empirically confirmed; there is firm direct evidence in the change with relative motion of the decay rate of mesons and in the relativistic Doppler shift.[3]) I will, however, suggest an interpretation wherein we can accomodate time-rate invariance for entropy clocks within the special relativity theory.

IV. The Interaction Interpretation

We recall that a particle or event is specified in space-time by a 4-vector (x, y, z, t), and that the conjugate momentum-energy is also given by a 4-vector (p_x, p_y, p_z, E), each component

[3]) These observations are described in detail in [10, chap. VI].

of which is proportional to the proper-time derivative of the corresponding space-time component. We expect, then, that any situation requiring the Lorentz transformation of the time t will also require transformation of the corresponding energy E, and conversely. We in fact do find that in the classic observations which have been made of the Lorentz time transformation there is just this parallelism: in the relativistic Doppler shift there is observed transformation both of frequency (time) and of energy; also in mass-energy transformations, frequency and energy transform together; and, for the relativistic meson-decay rate changes it can be shown that energy transformation of decay products, without explicit assumption of time transformation, does require the observed decay-rate changes [11]. On the other hand, in the reading of a hypothetical entropy clock, involving a thermodynamic system, there is no relativistic interaction between observing system and the clock process (the passage of gas from one container to another). The actual reading procedure might well involve interaction through electromagnetic waves for which we would expect a Lorentz transformation if there were a high velocity between clock and observer; but the presumption of a time transformation for the internal clock processes is not in an obvious accord with a parallelism between transformation of time and energy components of the 4-vectors.

The central thesis of the proposed interaction interpretation[4]) is that we should only expect a relativistic time transformation when there is a manifest associated energy transformation. Generally in the *hypothetical* clock transformations which have been discussed in relativity theory there is no expectation of such a concomitance; rather, the internal clock action is assumed to proceed for the observer without an associated relativistic physical interaction (and energy transformation). It is a provision of the interpretation, then, that the rates of macroscopic clocks, as, for example, the entropy clocks which we have discussed, will not be affected by motion relative to an observer. But, of course, we will expect time-rate changes for all processes which do involve relativistic energy transformation for the observer of the process; and, as noted, the established instances of relativistic time change are for processes of this kind.

V. The Constancy of Light Speed

Presumed relativistic changes in macroscopic clock times have the role of insuring that observers will find light to have the *in vacuo* speed c regardless of the velocity of the light source relative to the observer. It is quite possible, however, to maintain that constancy by considering the Lorentz transformations to apply directly to the space and time variables of light waves rather than to clocks and rods that are used in measurement of light speed.

In the customary interpretation, a light signal spreading (spherically, let us say) from a source is regarded as a single, invariant entity; then, by virtue of their different clock readings, as given by the Lorentz transformation, relatively moving observers find the signal to move with

[4]) Detailed discussion is given by R. Schlegel [11] and also [10, chap. VIII].

speed c in every direction, in spite of the relative motion between the observers. Their clocks and rods adjust, in accordance with the Lorentz transformations, so as to give the constant speed c. But, in the interaction interpretation the photons from the source are regarded as moving variously, so as to have a speed c relative to the interacting observer; that is, Lorentz transformations applied to the photons (or, to the electromagnetic field) give appropriate transformations to frequency ν and wavelength λ, such that $\nu'\lambda' = \nu\lambda = c$, where ν' and λ' are the transformed quantities. The ν' and λ' values obtained by any observer who transforms ν and λ for his coordinate system are, of course, just those values which are in fact observed in the relativistic Doppler shift. The transformations, then, are on the electromagnetic waves, or photons, which are assumed to interact with an observer only at the relative speed c. We lose the light signal as a unique spreading front, the same for all observers, and replace it by a field which may be regarded as a superposition of various ν and λ magnitudes, one set of which give a speed c relative to the observer and is actualized in the interaction with the observer. Obviously, there is still a relativistic time transformation, but it is on the light frequency and not on clock times. (We do, incidentally, at once solve the clock-paradox problem, since now there would be no relativistic aging differences, for clocks or for twins.)

VI. Empirical Studies

The interaction interpretation gives us a way in which an invariance of time rates for thermodynamic clocks, and for non-interacting entropic processes generally, can be brought into special relativity theory. The proposed new view involves little disturbance of the present mathematical structure of the theory, nor of its empirical foundation, since, as we have noted, experimental observations have generally been for consequences of physical interactions with parallel time and energy transformations. And yet, the interaction interpretation must of course be regarded as highly speculative, particularly in view of the many successes of the established interpretation of special relativity theory.

A test of the interaction point of view is provided by observations of relativistic rate changes of moving macroscopic clocks. Even without construction of a clock whose entropy can actually be determined, we may surmise that there should be no relativistic time change in a clock whose rate is a consequence of internal processes which are not in relativistic physical interaction with the observer. The cesium atomic clocks now in use are of this kind, and a first test of relativistic time effects has been carried out by J. C. Hafele and R. E. Keating, using portable cesium clocks in round-the-world jet aircraft flights [12]. Their results were in accord with the expected relativistic time changes. The clocks were carried around the world, once in an Eastward and once in a Westward direction. In the former case their motion was in addition to that of the axial rotation of the Earth, and in the latter it was opposed to the rotational motion. The net result was that, relative to a hypothetical non-rotating Earth which could be taken as an inertial system, the Eastward moving clocks ran more slowly than did the Westward clocks. Hence, Hafele and Keating found the clocks taken Eastward around the Earth to have lost time relative to rest clocks at Washington whereas those taken Westward gained time.

However, the round-the-world clock path is circular, and hence the clock motions in the Hafele-Keating experiment were accelerated. What was confirmed, therefore, was the general relativistic time effect. Specifically, the slight circular motion of the clock gives a centrifugal force that has the same effect as a gravitational potential Φ; and, the rate-monitoring photons in the cesium atomic clocks are therefore subjected to the general relativistic time-rate change factor, $\Delta\Phi/c^2$. The time changes calculated from this effect are the same, for the Hafele-Keating clocks, as one finds by using special relativistic equations.[5]) Since the clock motions were in fact accelerated, it seems proper to use the general relativity equations. The question of an *universal* time-rate effect, resulting only from *uniform* relative motion, then remains an open one empirically. It is hoped that further experimental studies will be made at an early date.

Tests of the universal application of the Lorentz transformation may be regarded as evidence for or against a constitutive role of a time entity in the natural processes of change. We may hope that in addition to solving problems of formal consistency we shall be profiting from further empirical observations of the relativistic properties of entropy. The second law of thermodynamics, expressed in terms of the entropy concept, has already given us a widespread objective correlate of the passage of time. It is quite possible that the investigation of relativistic transformations, in the domain of processes involving thermodynamic entropy, can be a probe for learning more of the place of time in our natural universe.

References

[1] M. von Laue: Die Relativitätstheorie—Erster Band: Die spezielle Relativitätstheorie. 5 Aufl. Vieweg & Sohn, Braunschweig 1952.

[2] R. C. Tolman: Relativity, Thermodynamics, and Cosmology. Oxford Univ. Press, Oxford 1934.

[3] H. Ott: Lorentz-Transformation der Wärme und der Temperatur. Zeits. f. Physik *175* (1963), 1, 70—104.

[4] L. A. Schmid: Effects of Heat Exchange on Relativistic Fluid Flow. In: A Critical Review of Thermodynamics (E. B. Stuart, B. Gal-Or, A. J. Brainerd, Eds.). Mono Book Corp., Baltimore 1970, 161—202.

[5] R. Schlegel: Time and Entropy. In [9], 27—35.

[6] L. Brillouin: The Arrow of Time. In [9], 101—110.

[7] I. Szumilewicz: The Direction of Time and Entropy. In [9], 181—192.

[8] J. Zeman: The Concept of Gradient Channel in Ontology and Epistemology and the Problem of Time. In [9], 193—207.

[9] Time in Science and Philosophy (J. Zeman, Ed.). Academia and Elsevier, Prague and Amsterdam 1971.

[10] R. Schlegel: Time and the Physical World. Dover, New York 1968.

[11] R. Schlegel: The Interaction Interpretation of Special Relativity Theory I and II. Foundation of Physics *3* (1973), 2, 169—184; 3, 277—295.

[12] J. C. Hafele and R. E. Keating: Around — the-World Atomic Clocks: Predicted Relativistic Time Gains. Science *177* (1972), 4044, 166—167. Around—the World Atomic Clocks: Observed Relativistic Time Gains. Science *177* (1972), 4044, 168—170.

[13] R. Schlegel: Flying Clocks and the Sagnac Effect. Nature *242* (1973), 5394, 180.

[5]) Details of the calculation are given in R. Schlegel [11, Part II]; a further aspect of the experiment is discussed in [13].

ENTROPY CONCEPTS IN PHYSICS

Michael J. Harrison

The concept of entropy was first introduced with reference to thermodynamic phenomena and their systematic explanation in the laws of thermal physics. Subsequent development and elaboration of the concept of entropy has revealed an extremely deep and powerful notion whose explanatory utility has extended far beyond the confines of the thermodynamic phenomena that initially gave rise to the idea. But it was in physics that the concept of entropy first began its role in science. In particular, those physical phenomena that result from the conjunction of mechanical and thermal agencies are the empirical sources of the concept of entropy. The interaction between mechanical and thermal phenomena provided the basic foundation for the concept of entropy, and also provided important intellectual stimulation for the further refinement of the entropy concept, both in science and engineering. There exists a systematic relation between the entropy of macroscopic objects and the measure of available microscopic information regarding them. This systematic relation has been generalized far beyond the original conditions of thermodynamic equilibrium first studied. Furthermore, the objects entering the systematic relation through which the concept of entropy is expressed need not be physical objects at all. The idea of entropy has been so generalized that it has found fruitful application in fields far removed from its birthplace. Not only in thermal physics, but in mathematics, information theory, biology, cosmology, chemistry, linguistics, communication theory, and philosophy the concept of entropy has revealed itself as a powerful analytical tool. But the historical roots of the concept are in the physics of macroscopic thermodynamic phenomena. The attempts to understand these phemomena that arose from engineering studies of heat engines in the eighteenth and nineteenth centuries provided the initial effort toward founding thermodynamics as a science. It is in these engineering studies that one may first discern the emergence of the concept of entropy in physics.

I. Historical Background

The science of thermodynamics developed in response to an accumulation of actual experience with heat engines. These engineering studies made it possible for empirical generalizations to be made regarding the behavior of heat engines, and such generalizations constituted the first formulations of the laws of thermodynamics.

James Watt invented the steam engine in 1765, and soon serious attention began to be given to the relations between thermal energy and mechanical energy [1, p. 8]. Subsequently, the experiments of Benjamin Thompson in 1798 established in a general way the equivalence of mechanical energy expended and thermal energy developed in the course of doing work against friction [1, pp. 8–9].

In 1840 James Prescott Joule established quantitatively the relation between thermal energy and mechanical energy by measuring the mechanical equivalent of heat [1, pp. 10–11]. Once the conversion factor relating the units of mechanical and thermal energy was known it was soon discovered and recognized that heat itself is a form of energy. The complete formulation of the relation between thermal and mechanical energy was given by Robert Julius Mayer in 1842 in his statement of the principle of conservation of energy [2, p. 493]. The content of Mayer's formulation is the first law of thermodynamics:

$$\mathrm{d}U = \text{đ}W + \text{đ}Q, \tag{1}$$

where $\mathrm{d}U$ is the increment in the internal energy of a system when an increment of heat $\text{đ}Q$ is transferred to the system and an increment of work $\text{đ}W$ is done on it.

The second law of thermodynamics is the empirical basis for the introduction of the entropy concept in physics. Initially, observations of Carnot in 1824 resulted in the statement of certain limitations in transforming heat into work using heat engine cycles [2, pp. 506–510]. Later, Rudolf Clausius in 1850 developed one alternative statement of the second law of thermodynamics: heat by itself cannot pass from a colder to a hotter body without causing other changes [1, pp. 16–17]. Lord Kelvin in 1851 formulated another alternative statement of the second law: a process whose only effect is the complete conversion of heat into work cannot occur [1, pp. 17–18]. There are still other macroscopic formulations of the second law of thermodynamics. Since it is an empirical law, based on actual experience with heat engines, all these alternative formulations are equally valid, and in fact are all equivalent to each other.

The first specific appearance of the entropy concept in scientific thought was associated with the development of thermal physics after the various formulations of the second law of thermodynamics had flowed from observations of phenomena manifested in the operation of heat engines. The entropy change ΔS occasioned in a system by the transfer to it of an amount of heat ΔQ was first introduced into physics by Rudolf Clausius in 1865 [3]:

$$\Delta S = \frac{\Delta Q}{T}, \tag{2}$$

where T is the absolute temperature of the system when the heat ΔQ is absorbed. Clausius studied the properties of the entropy S as defined above, and in his discussion of heat cycles he gave an abstract and general formulation of the second law of thermodynamics solely in terms of the properties of the system entropy S: – During real processes, the entropy S of an isolated system always increases. In the state of equilibrium the entropy attains maximum value.

Clausius general formulation of the second law in 1865 in terms of a mathematically defined quantity having a macroscopic basis represents a theoretical triumph of a very high order. The introduction of the entropy of a system through equation (2), or through its integral form

$$S_2 = \int_1^2 \frac{đQ}{T} + S_1, \tag{3}$$

provides the mathematical language for making a statement describing the direction of evolution for natural processes. The definition of entropy by Clausius allows us to restate the second law of thermodynamics in a form which requires all natural phenomena to proceed in the direction of increasing S. Since the second law then prescribes the direction of natural change it takes on very great significance in any philosophical discussion of time. The idea that time flows in one direction only is common to all direct human experience of the world. All macroscopic evidence, both ordinary and scientific, requires the conclusion that time passage is unidirectional. Such a conclusion is consistent with all biological and thermodynamic processes that we are aware of at the macroscopic level. The fundamental accomplishment embodied in Clausius' formulation of the second law is to attach the principle of increasing entropy as a signature to the direction of time flow that nature has adopted at the macroscopic level.

How and why nature has adopted this particular signature, specifying the necessary increase of entropy S in all natural macroscopic processes, is not easily perceived. The matter is far from transparent; all the more so since almost all of the fundamental laws of microscopic physics remain unchanged when the time variable t is replaced by $-t$. That is to say, these fundamental laws of physics at the microscopic level are invariant under time reversal, and allow any microscopic process to proceed in either direction of the time axis. All the fundamental laws of microscopic physics which are regarded as relevant to an understanding of macroscopic systems are time reversible; in these laws no unidirectional flow of time is indicated. It is correspondingly difficult to comprehend the second law of thermodynamics in the formulation in which entropy tends to increase inexorably with the passage of time. The time reversibility of the microscopic physical laws stands alongside the time irreversibility of the second law of thermodynamics. In fact it is the direction of change provided by an increase in entropy whick marks the forward flow of time, and indeed which provides a comprehensible foundation for the concept of the forward flow of time.

Classical thermodynamics is an abstract theoretical system which may be regarded as deduced from two fundamental postulates: the first law of thermodynamics, and the second law of thermodynamics in the formulation utilizing the entropy concept. Together with definitions of heat, work, and temperature, these postulates support a purely deductive science of thermodynamic phenomena. The concept of classical thermodynamics in no way refer to the microscopic structure of material systems, and in no way presume any underlying model to be used in calculating the values of the various thermodynamic quantities involved. The different quantities are regarded as phenomenological entities with mutual relations obtaining among them as prescribed by the first and second laws of thermodynamics. All the consequences of classical thermodynamics are then regarded as phenomenological statements mathematically deducible from the basic postulates.

The deductive character of classical thermodynamics perhaps received its most formally structured representation in the axiomatic approach of Constantine Carathéodory in 1909 [4, p. 355]. He developed an elegant axiomatic formulation of entropy and the second law in geometric terms. The geometrical language employed refers to the space of points representing the possible thermodynamic states of a system, with each such state specified by a complete set of thermodynamic coordinates. In Carathéodory's treatment the idea of heat is introduced as a secondary concept. The principle of conservation of energy is adopted ab initio for adiabatic changes in the state of a system. The first law is then formulated without explicit reference to the concept of heat, since for adiabatic changes no heat transfers occur and none need be mentioned. Adiabatic isolation and adiabatic changes of thermodynamic coordinates are adopted as undefined primitive concepts reflecting empirical observations. An axiom or postulate is then introduced which appears to represent an extremely succint starting point for the axiomatic development of thermodynamics. This postulate is known as "Carathéodory's Principle", and may be phrased as follows: "For a given thermal equilibrium state represented by a point in the space of thermodynamic coordinates, there exists in every neighborhood of the state-point another state-point which can never be reached from the first point by an adiabatic process." The utility of this postulate rests on its being able to lead to the concepts of entropy and absolute temperature, and ultimately to the introduction of the idea of heat. The development of the theory is based on the theorem from mathematics which asserts that a Pfaffian of the form

$$đQ = X\,\mathrm{d}x + Y\,\mathrm{d}y + Z\,\mathrm{d}z$$

has an integrating factor if, and only if, there exists in every neighborhood of a point P at least one point P' which cannot be reached from P along any curve on the surface determined by the condition $đQ = 0$.

The theorem quoted above essentially guarantees that if two equilibrium states are close to each other but not mutually accessible by an adiabatic process, then one can define an appropriate heat transfer $đQ$ and corresponding integrating factor $1/T$ so that

$$\mathrm{d}S = đQ/T$$

is an exact differential. In this way the concept of entropy, and the content of the second law, can be introduced through the route provided by Carathéodory's Principle. Although such a formal approach to the concept of entropy may not add any physical information distinct from that contained in the historically earlier formulations, it is useful to have as systematic a formulation of any physical theory as can be devised. There is value in the logical structure revealed by a thoroughly axiomatic theory developed with the aid of mathematical theorems and resting on a minimal set of assumptions. Such an exposition adds philosophic completeness as well as breadth of understanding to our comprehension of the basic physical relations described by the theory. Our conceptual grasp of entropy and indeed all of classical thermodynamics is enriched by the approach of Carathéodory. His axiomatic approach allows us to discern more clearly what are the basic phenomenological elements of the theory, and what is the purely logical component which binds those experimental facts together with mathematical force.

II. Microscopic and Statistical Concepts of Entropy

The concept of entropy introduced historically by Clausius is purely a macroscopic idea. The tendency for the entropy of an isolated system, or indeed of the whole universe, to increase in the natural course of events is an empirical generalization from macroscopic experience. By itself the second law of thermodynamics makes no reference to any microscopic causes or underlying structures which insure that the law necessarily follows. The source of the second law of thermodynamics cannot be found at the macroscopic level of experience. Any inquiry into the fundamental physical reasons underlying the law must be directed toward microscopic structures which provide a description of matter and motion on the scale of the smallest objects considered relevant to physical explanation.

The smallest material objects and the appropriate description of their motion have been variously considered in the course of the development of physics. The kinetic theory of gases constructed in the preceding century envisaged a collection of a great number of impenetrable atomic or molecular particles which have specific masses and move in accordance with classical laws of motion under mutually interacting forces. The modern transcription of the gas kinetic model involves the attribution of both quantum mechanical structure to the ponderable atoms and molecules, and motion according to the laws of quantum mechanics. Yet many features of the modern microscopic physical description of matter are conceptually similar to corresponding aspects of the older formulation. The concept of entropy received a microscopic description and explication in pre-quantum physics which shares many features in common with the modern microscopic version cast in the language of quantum mechanics. There are truly important differences in concept between the two formulations. And these differences lead to crucially distinct physical consequences in a number of important instances and thermodynamic circumstances. But the basic insight which provides a microscopic foundation for the concept of entropy in classical physics also serves in transcribed form as the basic insight underlying the definition of entropy in a quantum description of matter. For this reason there is great value in first discussing the historical attempts to provide a microscopic foundation for the concept of entropy. In fact, the analogy between the classical and quantum mechanical treatments of the concept of entropy is sufficiently useful to justify its very common use as a pedagogical tool.

It was Ludwig Boltzmann who provided the first microscopic definition of entropy. His definition was statistical in character and was based initially on the statistical mechanical description of a gas of molecules. In 1872 Boltzmann [5, pp. 18–25] developed his celebrated H Theorem for a gas of molecules described by a velocity distribution function $f(\vec{v})$. Boltzmann defined the quantity H given by:

$$H \equiv \iiint f(\vec{v}) \log f(\vec{v}) \, d\vec{v}, \tag{4}$$

where $f(\vec{v}) \, d\vec{v}$ is the number of molecules in the velocity space interval $d\vec{v}$. Using the laws of classical mechanics Boltzmann showed that under some very reasonable assumptions regarding the effects of molecular collisions the quantity H never increases.

Later, in 1877 [6, pp. 373], Boltzmann was able to identify the quantity H with the entropy S, to within a negative constant of proportionality; $-kH \equiv S$. Thus he obtained an explicit microscopic expression for the entropy which possessed the property that it never decreased:

$$S = -k \iiint f(\vec{v}) \log f(\vec{v}) \, d\vec{v}. \tag{5}$$

Further, it can be shown that the above definition for S leads to the result $dS = đQ/T$.

Subsequently Boltzmann generalized the gas kinetic definition of S [7] and arrived at a more general statistical mechanical definition of S valid for an arbitrary system obeying classical mechanics in canonical form:

$$S = -k \iint dp \, dq \, f(q. p) \log f(q, p). \tag{6}$$

This conceptual generalization provided Boltzmann with the vehicle to release statistical mechanics from its initial connection to the kinetic theory of gases and the more restricted methods of that latter discipline. His insight gave Boltzmann the means to construct a microscopic theory of thermodynamic phenomena for general systems of weakly interacting particles having potential energy arising from external force fields as well as kinetic energy. This theory was firmly based on the principles of probability theory conjoined with the laws of classical mechanics.

Later, Boltzmann's research was developed further by Planck [8, pp. 119–123] who extracted the statistical implications of Boltzmann's thought and summarized his definition of entropy S in a statistical relation equivalent to that given by Boltzmann:

$$S = k \log W. \tag{7}$$

In the above expression k is a numerical constant, and W is the number of physically distinct microscopic states of the system which correspond equally well to a given macroscopic state specified by known values of thermodynamic coordinates. The quantity W is often called the thermodynamic weight of a state which has been macroscopically characterized as having thermodynamic coordinates with given values. W is therefore the number of distinct microstates which agree with what is actually known about the state of the macroscopic system. The formula $S = k \log W$ for the statistical entropy of a system is essentially Boltzmann's definition of S. From this formula the first and second laws of thermodynamics are readily derived.

The outstanding achievement represented by the Boltzmann formula $S = k \log W$ is that it defines the entropy S absolutely. No ambiguous additive constant is present in this formula once the thermodynamic weight W is uniquely defined. Classical thermodynamics offers no such possibility for a unique and absolute definition of the value of entropy since only the entropy difference $dS = đQ/T$ is defined at the level of macroscopic thermodynamics. Only entropy differences between distinct equilibrium states are defined in classical thermodynamics. The Boltzmann formula $S = k \log W$ allows an absolute determination of the value of the entropy in any equilibrium state.

Boltzmann may be regarded as the first to have appreciated the role of statistical considerations in the concept of entropy. The expression $S = k \log W$ appears on Boltzmann's grave in the Central Cemetery of Vienna. His formula represents his great gift to our understanding of the physical universe and the observed directions of temporal change exhibited by natural processes wherever and whenever they occur.

The next great elaboration of the concept of entropy is due to the extraordinary efforts of J. W. Gibbs in developing statistical physics as a systematic science. Gibbs had formulated a probabilistic interpretation of entropy much earlier, but in 1902 he summarized his thinking on the subject in a book: "Elementary Principles in Statistical Mechanics" [9]. The approach employed by Gibbs introduced the use of general ensembles to represent systems of actual interest. A general ensemble in this context is taken to be a collection of mental copies of the system of actual interest. The copies are intended to be exact replicas of the actual system, and they are regarded as distributed among all the available microsocpic states consistent with the thermodynamic coordinates characterizing the system of actual interest.

In classical mechanics a microscopic state of a system having N degrees of freedom corresponding to N configuration coordinates is specified by a point in a $2N$ dimensional phase space representing the coordinates and their canonical momenta. The phase space is often denoted as Γ space. Gibbs gave the first systematic foundation of a general theory of statistical mechanics in terms of Γ space. In his formulation a Γ space appropriate to the system of actual interest was regarded as populated with moving phase points representing the microscopic states successively attained by the various virtual copies of the real system. These virtual and exact copies of the real system collectively constitute an ensemble, in the sense described above. Thus, Γ space contains phase points representing the states instantaneously occupied by the members of the ensemble; and there are as many phase points moving around in Γ space as there are members of the ensemble.

These phase points are statistically assigned to different regions of Γ space according to certain principles. Thus, the phase points are distributed in Γ space in a manner consistent with available partial knowledge of the mechanical state of the system of actual interest, and in a manner which accords with certain a priori probabilities which are themselves consistent with that partial knowledge. Once the distribution of phase points representing the members of the ensemble has been introduced in Γ space, the entropy may be defined as a functional of the phase point distribution. Therefore, in the formulation of Gibbs the entropy is regarded as a statistical concept resting on the idea of a representative ensemble. Entropy is conceived then primarily as an ensemble property. Since the given ensemble has been chosen to reflect fully whatever information and partial knowledge we may have of the system of actual interest, it is evident that an ensemble theoretic definition of entropy provides a link between information and entropy.

In applying the method of Gibbs the kind of representative ensemble chosen will depend on what sort of information really characterizes the system of actual interest. The ensemble chosen will itself in turn determine the value of the thermodynamic weight W, which enters the Boltzmann formula for the entropy $S = k \log W$. The calculation of W depends directly on the ensemble chosen, and in general will be different for different ensembles. In particular, the value of W tends actually to increase in time as the appropriate representative ensemble itself evolves in parallel with whatever natural processes are occurring in the macroscopic system of actual interest.

The Boltzmann formula $S = k \log W$ most usefully represents the entropy of a statistical ensemble in equilibrium; that is one with time independent properties. Such an equilibrium ensemble describes theoretically a macroscopic system in thermodynamic equilibrium. It is possible to define the entropy S alternatively for an arbitrary equilibrium statistical ensemble. The method is suggested by Boltzmann's H Theorem, but was stated more precisely by Gibbs in 1902:

$$S = -k \int \mathrm{d}\Gamma\, f(q, p) \log C f(q, p) \equiv -k \langle \log Cf \rangle_f. \tag{8}$$

In the expressions above the function $f(q, p)$ represents the normalized probability density in Γ space for finding a phase point representing an ensemble member, and the brackets $\langle\,\rangle_f$ denote an average over this probability density distribution $f(q, p)$. The quantity C is some constant chosen to describe a cell size in Γ space which classically embodies a set of microscopic states sufficiently similar to be regarded as experimentally alike as far as macroscopic measurements are concerned. Thus, the size of the constant C determines how fine or coarse a granular structure has been introduced into Γ space by the intrinsic limitations of the information gathering experimental apparatus that is used to make measurements. A small value of C implies a fine-grained division of phase space, and a large value of C implies a coarse-grained division of phase space corresponding to less sensitive information gathering and less detailed empirical observations. The value of C therefore provides a link between the detailed division of phase space into cells of a certain size, and the limitations on fineness of detail that necessarily accompany the use of any actual experimental apparatus.

It can be proved that the above expression in Eqn. (8) provides a more general definition of the entropy S which becomes useful also in the case of time dependence. One has only to regard the distribution function f as a function of time t as well as the canonically conjugate coordinates and momenta: $f = f(q, p, t)$. Such a formulation naturally describes an ensemble which itself is changing in time. A time-dependent ensemble must be employed to represent whatever time-dependent partial knowledge is available about a system of actual interest. This more general definition of S provided by Eqn. (8) is precisely equivalent to $S = k \log W$ for any special case represented by an equilibrium ensemble. In particular, the definition of S based on Eqn. (8) gives correct expressions for the statistical entropy in the three cases of a microcanonical, canonical, or grand canonical ensemble, which respectively describe a system of fixed energy and particle number, fixed temperature and particle number, or merely fixed temperature and average particle number.

The virtue of the expression

$$S = -k \int \mathrm{d}\Gamma\, f \log Cf$$

is that it therefore readily generalizes the concept of entropy to time-dependent situations, where a time-dependent macroscopic system is not in equilibrium and is appropriately represented by an explictly time-dependent ensemble distribution function $f(q, p, t)$.

It should be noted that the definition $S = -k\langle \log Cf \rangle$ for the entropy depends on the choice of the constant C as far as the absolute value of the entropy is concerned. But entropy differences between any two macroscopic states are completely independent of the choice of C and unaffected by the numerical value of C. Since only entropy differences between equilibrium states are introduced and discussed in classical thermodynamics, any choice of C in Eqn. (8) is consistent with the principles of classical thermodynamics. But just as the latter possesses a concept of entropy which is determined only to within an arbitrary additive constant, so does Eqn. (8) for the statistically defined entropy leave its numerical value undetermined to within an additive constant which depends on C. There has nevertheless been a great advance in our understanding of the concept of entropy in going from the realm of classical thermodynamics to the microscopic description provided by statistical mechanics. For, to realize that the source of the arbitrary and undetermined additive constant in the entropy is the presence of the constant C in Eqn. (8) is to realize an important conceptual link between classical thermodynamics and the statistical description of nature at the microscopic level. Furthermore, when the laws of classical mechanics are replaced by the principles of quantum mechanics the physical description of a material system is sufficiently altered to remove any arbitrary additive constant entering the statistical definition of entropy. Thus, when quantum mechanical principles are introduced the microscopic concept of entropy becomes absolutely precise and free from any arbitrary elements arising from phenomenological considerations. As we shall see this uniqueness in the numerical value of the statistical entropy brought about by the introduction of quantum mechanics leads to an important new law of thermodynamics.

The transition to quantum mechanics from classical mechanics takes a straightforward path. The result of employing quantum mechanics to describe the motion of material particles leads to the primary and fundamental nature of quantum statistical mechanics.

The first point to note is that in quantum mechanics the microscopic state of a system is specified by a wave function rather than by a set of canonically conjugate coordinates and momenta. In the Schrödinger picture of quantum mechanics the passage of time is marked by a time dependence of the wave function.

The notion of ensembles is also employed in quantum statistical mechanics to represent a system of actual interest. But such ensembles are appropriately defined in quantum mechanical terms. These quantum mechanical ensembles are then constructed as virtual assemblies of similar systems possessing identical Hamiltonian operators, any one of which gives an accurate dynamical description of the system of actual interest.

As before in classical statistical mechanics, so again in quantum statistical mechanics the condition of macroscopic thermodynamic equilibrium requires the use of an equilibrium ensemble to represent the system. This necessity to employ an equilibrium ensemble leads to the requirement that the stationary eigenstates of the Hamiltonian for the system be adopted as constituting the language in which the description of the system wave function is to be given. That is to say, an equilibrium ensemble implies that the energy eigenstates having definite energy eigenvalues associated with them are the appropriate states on which to base a description of the wave function of the system. Such an appropriate set of states is called a representation: macroscopic equilibrium implies the choice of an energy representation. Thus we see that the foregoing leads to the necessity of adopting the energy representation for the description of whatever quantum mechanical information we may have about an actual system in a condition of thermodynamic equilibrium.

In such a quantum mechanical description as is provided by the Hamiltonian representation the energy eigenstates Ψ_l having energy levels ε_l are occupied with probability amplitudes a_l and ensemble average occupancy probabilities $P_l \equiv \overline{|a_l|^2}$, where the bar above $|a_l|^2$ denotes an average over the members of the ensemble.

Whatever macroscopic information is available must be related to the state (wave function Ψ) characterizing the macroscopic system of actual interest. Such information must be quantitatively embodied in an ensemble theoretic description yielding statistical information about the phase and amplitude of the wave function of the system. In the energy representation generated by the eigenstates Ψ_l of the Hamiltonian this means knowledge of the phase and amplitude of the expansion coefficients $a_l(t)$ appearing in

$$\Psi = \sum_l a_l(t)\, \Psi_l,$$

where we have noted previously

$$\overline{|a_l|^2} = P_l.$$

In general $a_l(t)$ depend on time for each member of the ensemble, and the ensemble average value of $|a_l|^2$ provides the occupancy probabilities P_l.

In order to complete the prescription for constructing an appropriate quantum mechanical ensemble in any given case it is necessary to introduce certain postulates and a priori assumptions regarding the incorporation of whatever partial knowledge we may have of the system of actual interest. We must decide how to link the information available about the system with the quantum mechanical language employed in an ensemble theoretic description. To this end there are adopted the twin postulates of equal a priori amplitudes $|a_l|$ and random phases for states Ψ_l of equal a priori probability, consistent with whatever information is available about the system. Thus, the equal a priori amplitudes and random phases are postulated in a context which requires overall consistency with what is actually known about the extent to which the states Ψ_l are occupied by the system of actual interest.

One can again employ microcanonical, canonical, or grand canonical ensembles, defined similarly as in classical statistical mechanics. Such equilibrium ensembles again serve as useful ensembles to represent systems in thermodynamic equilibrium. In both quantum and classical statistical mechanics the equilibrium state is characterized by a further postulate which takes account of the perception that a system which is either isolated or subject to fixed values of externally controlled thermodynamic variables will approach equilibrium as it approaches a condition of maximum probability consistent with the constraints upon it. The description of the constraints is carried through in the ensemble theoretic language employed to express whatever partial knowledge is available. The statistical description of the condition of maximum probability consistent with the constraints will depend on whether quantum or classical statistics is being employed, and on what sort of ensemble has been chosen. But in all cases the equilibrium state is defined by a postulate which identifies it with the ensemble theoretic distribution having a maximum probability consistent with the constraints provided by partial information. Thus, by definition the equilibrium state is the condition of greatest probability, given whatever constraints are present.

The concept of entropy in quantum statistical mechanics is introduced by analogy with the formulations of Boltzmann and Gibbs in classical statistical mechanics. Given the occupancy probabilities P_l, the entropy is then defined according to the expression below:

$$S \equiv -k \sum_l P_l \log P_l \tag{10}$$

$$= -k \overline{\log P_l}. \tag{11}$$

In Eqn. (11) the bar above $\log P_l$ refers to an ensemble average using the P_l as the probability distribution over the energy eigenstates.

Equivalently, one can develop Eqn. (10) into a quantum mechanical statement of the Boltzmann formula for the statistical entropy of a system in thermodynamic equilibrium:

$$S = k \log W(U, \bar{N}, V, \bar{\alpha}). \tag{12}$$

In Eqn. (12) above the thermodynamic weight W is now defined to be the number of microscopically specified possible quantum states corresponding to the given macroscopic equilibrium state of the system. The equilibrium state is regarded as characterized by the ensemble average values $U = \bar{E}$, $\bar{N}$, $\bar{\alpha}$, and V for the values of internal energy, average particle number, average values of other internal state parameters, and volume, respectively. The macroscopic thermodynamic state is specified then by U, $\bar{N}$, $\bar{\alpha}$, and V.

Given Eqn. (12) for the entropy the absolute temperature T may then be defined by:

$$\frac{1}{T} \equiv \frac{\partial S}{\partial U}. \tag{13}$$

Eqn. (12) and Eqn. (13) provide a conceptual foundation for a complete statistical mechanical derivation of the laws of classical thermodynamics. Thus, they lead to a microscopic and statistical understanding of all thermal phenomena observed at the level of macroscopic physics.

The conceptual relation between entropy and probability provides a probabilistic interpretation of the second law of thermodynamics as formulated by Clausius. Let S_{tot} denote the total entropy of an isolated and very large system (the universe as a possible example), and let this large system consist of a minor part having entropy S and a major part having entropy S_e. The major part is regarded as constituting the external environment of the minor part. Then $S_{tot} = S + S_e$, and Clausius' statement of the second law asserts that

$$\mathrm{d}S_{tot} \geqq 0 \tag{14}$$

in any process, with the equal sign holding only for a reversible process occurring in the total system having entropy S_{tot}. For an external environment undergoing a quasi-static change at temperature $T_e = T$ we have

$$dS_e = -đQ/T, \tag{15}$$

and for the system in thermal contact with the environment we therefore have

$$dS \geqq đQ/T. \tag{16}$$

The probabilistic interpretation of the second law follows from linking Eqn. (12) for the statistical entropy with Eqn. (14) which states that the total entropy of an isolated system must increase or stay the same in any natural process. Since Eqn. (12) bases the value of the entropy S on the value of the thermodynamic weight W in a monotonic increasing relation it can be seen that macroscopic states of higher entropy correspond to thermodynamic state coordinates which specify larger values of the thermodynamic weight W. But W is the number of distinct microscopic states which are equally well consistent with the available knowledge describing the macroscopic thermodynamic state. W is therefore proportional to the probability of finding the system in the given macroscopic state, since any such state which can be realized in a great number of microscopic ways is more probable than another state which can only be realized in few microscopic ways. Thus we conclude that the entropy of a thermodynamic state is a monotonic function of the probability of that state. And Eqn. (14) then asserts that natural processes move systems into thermodynamic states of progressively higher probability. In other words, the concept of entropy is consistent with the notion that left to themselves all systems tend to evolve into states of sequentially higher probability. The more probable tends to develop out of the less probable. And states of higher entropy tend to develop from states having less entropy.

On the average, the more probable will also occur with greater frequency than the less probable. But this property of more probable events is true only on the average. Less probable events can and do indeed occur, but their incidence of occurrence is much less. Less probable events can be perceived with much lower frequency than more probable events. The occurrence of less probable events represents an improbable departure from what can normally be expected.

From the above discussion of the nature of probability it can be concluded that the tendency of the statistical entropy to increase is a probabilistic tendency. The entropy on the average will almost always increase, but rare occasional departures of small amplitude are possible in principle. Any spontaneous decrease of the entropy, defined by $S = k \log W$, would necessarily correspond to some highly improbable fluctuation in either the energy E, the particle number N, the volume V, or in other externally controlled parameters α. Such an improbable fluctuation would have to occur as a deviation from the average values of these quantities. It is the average values which serve as the thermodynamic coordinates of the system. The equilibrium value of the entropy, toward which the system returns after any improbable fluctuation, depends on these macroscopically defined average values. Temporary departures from equilibrium, in the form of fluctuations, represent conditions of lesser probability than the equilibrium condition. A spontaneous decrease in the entropy of any isolated system represents an extremely improbable departure of the entropy from its maximum equilibrium value.

Thus the second law of thermodynamics has a statistical significance. It is true on the average as a statement describing the overwhelmingly likely behavior to be expected from any randomly selected member of the ensemble chosen to represent the actual system. The most probable behavior of the system represents its average behavior. And the ensemble average state is a good representation of the most probable state. Fluctuations about the most probable state can occur, but large fluctuations are most improbable. Furthermore for large systems having many degrees of freedom the most probable values of physical quantities coincide with their average values because the distribution of values is very sharply peaked about the most probable values in equilibrium.

The law of entropy increase for an isolated system is best understood as a statement about the time average behavior of the most probable state of a system with many degrees of freedom. In quantum statistical mechanics one can provide the following general definition of entropy S in the case of time-dependent phenomena:

$$S = -k \sum_l P_l(t) \log P_l(t), \tag{17}$$

where $P_l(t)$ are the time-dependent occupancy probabilities. In equilibium, when $\mathrm{d}P_l/\mathrm{d}t = 0$, the above definition reduces to the equilibrium state definition, $S = k \log W$. This is proved as follows:

In thermodynamic equilibrium one has

$$P_l = 1/W \tag{18}$$

for any state l in the group of W quantum states, and

$$P_l = 0 \tag{19}$$

if the state l is outside the group of W quantum states. Thus, Eqn. (17) leads to the result:

$$S = -k \sum_l (1/W) \log (1/W) \tag{20}$$

$$= +k(1/W) \log W \sum_l 1. \tag{21}$$

But by definition of the enumeration over l one has:

$$\sum_l 1 = W.$$

Therefore Eqn. (21) reduces to the equilibrium definition $S = k \log W$.

Similar probabilistic and statistical interpretations of the second law of thermodynamics hold both in classical statistical mechanics and quantum statistical mechanics. Our discussion above has had adequate generality to cover both cases. Each approach provides essentially the same broad understanding of the second law as formulated by Clausius in terms of the tendency for the entropy to increase: $\mathrm{d}S_{\mathrm{tot}} \geqq 0$.

There is however an important aspect of the concept of entropy in quantum statistical mechanics which has no analogue in classical statistical mechanics. The idea of entropy in physics is correspondingly more revealing and determinate in quantum statistical mechanics.

The additional aspect which arises from the use of quantum mechanics concerns the limiting value of the entropy as the absolute zero of temperture is approached. The constancy of this zero temperature value of the entropy, *independent* of the values of any and all other thermodynamic coordinates specifying the state of the system, is established through the use of quantum mechanical principles. And only through quantum mechanics is it possible to obtain the result that the zero temperature limit of the entropy is a constant which is independent of the particular thermodynamic state the system is in. This constancy of the entropy value in the limit of absolute zero temperature has come to be known as the third law of thermodynamics. The constant value of the entropy at absolute zero may be conveniently chosen as zero. The third law may then be formulated in its most commonly quoted form: the entropy per unit mass of any substance goes to zero as the temperature goes to absolute zero.

The quantum mechanical basis for the third law of thermodynamics may be readily perceived. In the quantum mechanical interpretation the formula $S = k \log W$ has no arbitrary constant whatsoever. The quantity W has its possible values completely well defined by the energy level structure of the system, and the Boltzmann formula defines the entropy uniquely. This was *not* the case in classical statistical mechanics. There one encountered the necessity for introducing some arbitrary phase space cell whose size was chosen to contain a set of microscopic classical states similar enough to be grouped together. The size of the arbitrary phase space cell is directly related to the choice of the constant C in Eqn. (8).

The unique value attached to the entropy by the formula $S = k \log W$ in quantum statistical mechanics arises from the well defined degeneracy of energy levels in quantum mechanics. Each energy level ε_l may have many distinct quantum states associated with it as eigenstates of the Hamiltonian operator describing the system. Any such multiplicity of quantum states associated with a given energy level is called a degeneracy in the energy level structure. It is a property of large macroscopic systems having many degrees of freedom that the degeneracy of the energy levels increases very rapidly with the energy eigenvalue. In this context the ground state can then be regarded as non-degenerate or as consisting of very few close lying levels when compared with the degeneracy of any state of higher energy eigenvalue likely to be populated at elevated temperatures. Furthermore, such non-degeneracy of the ground state will persist as externally controlled thermodynamic variables, such as volume, are changed from one value to another. The ground state remains non-degenerate. The Boltzmann formula $S = k \log W$ can be readily evaluated in the limit of absolute zero temperature by noting that only the ground state of a system can possibly be populated or occupied at absolute zero. Thus, for any system at absolute zero the thermodynamic weight W is necessarily given by the non-degenerate value $W_0 = 1$ referring to the ground state. Thus we have the following result as a consequence of quantum mechanics:

$$\begin{aligned} \lim_{T\to 0} S &= k \lim_{T\to 0} \log W \\ &= k \log W_0 \\ &= k \log 1 \\ &= 0. \end{aligned} \tag{22}$$

Therefore, for any macroscopic system we have the general result that the entropy vanishes at absolute zero:

$$\lim_{T \to 0} S(T) = 0, \tag{23}$$

independent of the values taken on by any external parameters whose variation causes work to be done on the system. Eqn. (23) is the mathematical statement of the third law of thermodynamics.

III. Modern Concepts of Entropy in Physics

The significance of the concept of entropy has been enriched in recent decades by developments in several directions. Some workers have generalized the concepts and principles of classical equilibrium thermodynamics so that the resulting theory represents a phenomological description of steady-state irreversible thermodynamics. Parallel theoretical efforts have been successfully made to provide a microscopic and statistical derivation of the phenomenological principles which constitute the theory of irreversible processes. All such treatments of irreversible phenomena are valid only near equilibrium. The thermodynamic processes comprehended by these theories occur when the system is disturbed from equilibrium by only a small amount. The departure from equilibrium is assumed to be sufficiently small for any irreversible flow of matter, thermal energy, or electricity, to depend at most linearly on the forces giving rise to non-equilibrium conditions. The central figure in the development of this theory of irreversible thermodynamics has been L. Onsager [10, p. 405; 11, p. 2265; 12].

The mathematical elements which enter the general theory of irreversible processes are linear equations relating various flow quantities with the force variables that give rise to these flows. Thus the theory describes the thermodynamics of transport phenomena in the regime where linear responses to external stimuli constitute an adequate description. The concept of entropy which enters the theory is extended beyond the equilibrium notion to include the ideas of entropy production in time, entropy density, and directed entropy flow. In the theory of irreversible processes entropy is therefore regarded rather like a substance, which can be created or destroyed, distributed throughout the system, and flow along various directions within the system. These properties of entropy flow are described by an entropy balance equation, which resembles an equation of continuity with a source term:

$$\partial \varrho s / \partial t + \vec{\nabla} \,.\, \vec{J}_s = \sigma, \tag{24}$$

where ϱ is the mass density per unit volume, s is the entropy per unit mass, ϱs is the entropy density per unit volume, σ is the entropy production per unit volume per second, and $\vec{J}_s$ is the entropy flux vector at each point of the system.

The generalized force variables that disturb the system from equilibrium and give rise to irreversible transport are denoted by X_i. The generalized fluxes which describe the transport are denoted by J_i. It can then be shown that the entropy production σ is given by a bilinear form in the forces and fluxes:

$$\sigma = \sum_i J_i X_i. \tag{25}$$

Furthermore, these fluxes J_i are linearly related to the forces X_i:

$$J_i = \sum_i L_{ij} X_j, \tag{26}$$

where the L_{ij} are an appropriate set of transport coefficients.

Consider now a system which is isolated both materially and thermally, so that neither matter, electricity, nor heat enters or leaves the system. Then in terms of the quantities defined above the entropy S_{I} associated with such a system changes with a rate given by:

$$\mathrm{d}S_{\mathrm{I}}/\mathrm{d}t = \int_V \sigma \, \mathrm{d}v \tag{27}$$

In general a system need not be thermally, or even materially, isolated from its environment. But in all cases the total entropy S of the universe will be the sum of the entropy S_{I} of the system of interest and the entropy S_{E} of the complete environment surrounding the system. Thus in general any change in the total entropy S of the universe is given by:

$$\mathrm{d}S = \mathrm{d}S_{\mathrm{E}} + \mathrm{d}S_{\mathrm{I}}. \tag{28}$$

Then for a system which is materially closed, but thermally open, the integral form of the entropy balance equation becomes:

$$\mathrm{d}S/\mathrm{d}t - \frac{1}{T} \,đQ/\mathrm{d}t = \int_V \sigma \, \mathrm{d}v. \tag{29}$$

The above equation represents the formulation of the second law of thermodynamics in the theory of irreversible processes for a closed material system to which heat is added at the rate $đQ/\mathrm{d}t$. This differs from the familiar statement $\mathrm{d}S \geqq đQ/T$ because the time t appears.

According to the usual statement of the second law, for any materially and thermally isolated system which remains closed one has the relation $\mathrm{d}S_{\mathrm{I}} \geqq 0$. From Eqn. (27) this then implies that $\sigma \geqq 0$ is the local form of $\mathrm{d}S_{\mathrm{I}} \geqq 0$. In the further construction of the theory of irreversible processes an important generalization is made regarding the function σ. We generalize the local relation $\sigma \geqq 0$ to open systems to obtain the basic postulate of irreversible thermodynamics as a general theory applicable to all systems near thermodynamic equilibrium. Thus we postulate that

$$\sigma \geqq 0 \tag{30}$$

and

$$\mathrm{d}S_{\mathrm{I}} \equiv \mathrm{d}t \int_V \sigma \, \mathrm{d}v \geqq 0 \tag{31}$$

are valid relations for open as well as closed systems, where $\mathrm{d}S_{\mathrm{I}}$ retains its definition as that part of the system's entropy change which arises from internal processes producing entropy inside the system.

A remarkable relation among the transport coefficients L_{ij} has been established by Onsager. He showed that when no magnetic field is present the coefficients obey the relation $L_{ij} = L_{ji}$. This relation is known as the Onsager reciprocity theorem, and these relations were proved by him on the basis of the principle of microscopic reversibility. The Onsager reciprocity relations can be regarded as phenomenological expressions of microscopic reversibility, and the starting point of a phenomenological theory of irreversible processes.

The probabilistic interpretation of the law of entropy increase as embodied in the second law of thermodynamics has been well established by empirical observations. Furthermore, considerable theoretical effort has gone into elaborating our understanding of the statistical basis of the law. However, deeper examination of questions related to the physical nature and origin of the second law reveals serious gaps in our understanding, and difficulties which remain unresolved.

One important difficulty relates to the application of the second law to the universe as a whole. It is clear both from everyday experience and astronomical observations that the universe as a whole is not in a state of thermodynamic equilibrium. And yet as a single closed system one might expect it to have reached equilibrium long ago. The temptation to regard the observable part of the universe as a very large fluctuation in a system in equilibrium as a whole can be resisted by realizing that comparable non-equilibrium conditions supporting biological development and human observers can occur on a smaller astronomical scale with even greater likelihood. There is a greater probability for a suitable fluctuation in the structure of the solar system since it contains far fewer degrees of freedom than the galaxy in which it is located [13, p. 29].

Landau and Lifshitz have sought the answer to this puzzle of cosmic irreversibility in the general theory of relativity [13, pp. 29–30]. They point out that large regions of space involve important considerations of the role of gravitational fields. The general theory of relativity describes gravitational fields as distortions of the space-time metric. Statistical physics ought then to regard such changes in the space-time metric as external conditions bearing on the thermodynamic behavior of matter. Clearly, if such external conditions remained constant over a long period of time, then the universe regarded as a closed system must eventually reach a state of equilibrium. But the space-time metric depends not only on position, but on time as well. Thus the external conditions provided by gravitational phenomena are not constant. In this explication of the puzzle the gravitational field is not itself counted as part of the closed system, and indeed the universe as a whole is regarded not as a closed system, but as one situated in a variable gravitational field. Thus the law of increase of entropy would not imply the necessity of statistical equlibrium. This problem of understanding the thermodynamic behavior of the universe as a whole is far from resolved. There are apparent contradictions, and much remains open to speculation.

Another recent discussion of fundamental issues arising in our efforts to understand the cosmological significance of the foundations of thermodynamics and time asymmetries has been given by Gal-Or [14]. He has raised the question explicitly of whether the origin of irreversibility in nature is local or cosmological, and whether the answers are to be found in the laws or the boundary conditions. In particular, Gal-Or raises the question of whether cosmology generates thermodynamics in the sense that the origin of the second law of thermodynamics can be traced back to the present expansion of the universe. The theory of general relativity plays an important role in such large-scale considerations of non-equilibrium dynamics in an expanding universe. Another basic approach to fundamental cosmological questions of irreversibility is sought in a union of the quantum principle and the theory of general relativity. These dual formulations are regarded as the two overarching principles of nature on which any understanding of cosmological irreversibility must depend.

All such attempts to broaden the concept of entropy in physics by introducing cosmological considerations which may or may not include boundary conditions must be regarded as tentative at the present time. The cosmological arrow of time which represents the observed irreversibility of nature is not yet fully understood. The origin of the asymmetry of time at the scale of the very large is not yet clearly perceived. And the relation between irreversibility in small systems and irreversibility in the very large system has not yet been fully discussed. Much more theoretical work as well as astrophysical observations will have to be performed before physics can speak articulately on these questions.

There has been another approach taken in recent decades to understand the concepts of entropy and irreversibility. These have been attempts to understand irreversibility at the microscopic level by careful systematic derivations of equations of motion exhibiting irreversibility. Such work begins only from quantum mechanical dynamics, and then proceeds to make well-defined mathematical approximations at each stage. The first theory of this genre was due to W. Pauli, who derived an equation of motion for the microscopic occupancy probabilities, P_l, assuming that the system wave function repeatedly suffered a random change of phase after *every* small time interval over which motion occurred [15, p. 30]. Such a repeated random phase approximation was shown to be unnecessary by L. Van Hove, who showed that only an *initial* random phase approximation was necessary provided the perturbation forces giving rise to interactions within the system had certain rather special, but ubiquitous, mathematical properties [16, p. 517; 17, p. 441; 18, p. 268; 19, p. 418]. The discussion of Van Hove is quite mathematical. We will content ourselves with the observation that his systematic derivation sheds further light on the concept of entropy in physics and refines our understanding of irreversibility accordingly.

Related fundamental discussions of the microscopic basis of irreversibility have been given by others, notably N. Bogolubov [20, p. 265] and I. Prigogine [21; 22]. The latter author has also extended the microscopic theory of non-equilibrium thermodynamics to encompass processes of biological change and evolution in living systems [23, p. 23]. The efforts of Bogolubov and Prigogine to provide a careful and rigorous derivation of irreversibility which starts from microscopic system dynamics represent a major advance in our physical understanding of the concept of entropy. The theoretical tools employed by them include techniques for analyzing non-linear phenomena, which enter their treatment of irreversibility in an essential way. There is currently much activity in this area which seeks to provide still firmer foundations at the microscopic level for the second law of thermodynamics. The concepts of entropy in physics, and related concepts concerning the temporal evolution of natural processes in macroscopic systems have a long history which continues to unfold.

References

[1] A. W. Porter: Thermodynamics. Methuen & Co. Ltd., 1946.

[2] G. Joos: Theoretical Physics. Hafner Publishing Co., 1950.

[3] H. Zeise: Thermodynamik. Hirzel, Leipzig 1944.

[4] C. Carathéodory; Math. Ann. *67* (1909), 355—386.

[5] D. ter Haar: Elements of Statistical Mechanics. Rinehart & Company, Inc., New York 1954.

[6] L. Boltzmann: Über die Beziehungen zwischen dem II. Hauptsatz der mechanischen Wärmetheorie und der Wahrscheinlichkeitsrechnung. Wein. Ber. *77* (1877), 373.

[7] L. Boltzmann: Vorlesungen über Gastheorie, Bd. 1. Barth, Leipzig 1923.

[8] M. Planck: Vorlesungen über die Theorie der Wärmestrahlung. Barth, Leipzig 1923.

[9] J. W. Gibbs: Elementary Principles of Statistical Mechanics. Yale U. Press, New Haven 1902.

[10] L. Onsager; Phys. Rev. *37* (1931), 4, 405—426.

[11] L. Onsager; Phys. Rev *38* (1931), 12, 2265—2279.

[12] S. R. de Groot: Thermodynamics of Irreversible Processes. North-Holland Publ. Co., Amsterdam 1952.

[13] L. D. Landau and E. M. Lifshitz: Statistical Physics. Pergamon Press Ltd., London 1958.

[14] B. Gal-Or: Entropy, Fallacy, and the Origin of Irreversibility. Ann. N. Y. Acad. Sciences *196* (1972), 6, 305—325.

[15] W. Pauli: Festschrift zum 60. Geburtstage A. Sommerfelds. Hirzel, Leipzig 1928.

[16] L. Van Hove; Physica *21* (1955), 517—540.

[17] L. Van Hove; Physica *23* (1957), 441—480.

[18] L. Van Hove; Physica *25* (1959), 268—276.

[19] L. Van Hove and E. Verboven; Physica *27* (1961), 418—432.

[20] N. N. Bogolubov; Journ. Phys. U.S.S.R. *10* (1946), 3, 265—274. Problems of a dynamical theory in statistical physics (transl. E. Gora). Providence College, Providence R. I. 1959.

[21] I. Prigogine: Etude thermodynamique des Phenomènes Irreversibles. Desoer, Liege 1947.

[22] P. Glansdorff, I. Prigogine: Thermodynamic Theory of Structure, Stability and Fluctuations. Wiley, New York 1971.

[23] I. Prigogine, G. Nicolis, and A. Babloyantz: Thermodynamics of evolution. Physics Today *25* (1972), 11, 23—28.

ENTROPY AND INFORMATION: CONFORMITIES AND CONTROVERSIES*

Johannes Peters

> *We find that the theories of physicists constantly undergo modification, so that no prudent man of science would expect any physical theory to be quite unchanged a hundred years hence. But when theories change, the alteration usually has only a small effect so far as observable phenomena are concerned.* [1, p. 213.]
>
> *Bertrand Russell*

1. How to determine what entropy is

In the course of time, notions and their designations emerge and change; consistency sometimes is preserved within a restricted field only. In different fields of science the same word may have different meanings so that context is necessary to exclude ambiguities.

When trying to learn what entropy is, we may look at the textbooks for verbal definitions. If this does not satisfy us completely, we may invent measuring devices where needle-deflections are to indicate observable facts which permit us to get this quantity by inference. We will not be satisfied unless there is a complete coverage within science which is unequivocal.

Unfortunately, this search for the meaning of entropy has to be extended over an interval of time lasting about one and a half centuries; it has to comprise many fields of physics and of technology. The intention here is, however, not to write a history of the development of the notion entropy in its different meanings; the intention is to investigate the intrinsic interconnections and to introduce suggestions about interpretations and designations with a tendency to eliminate controversies.

This work requires some restriction and organization from the very beginning thereby leaving aside many aspects which are by no means less important. By such an arbitrary decision we want to consider one extremely simple object in three fields of science. This is in so far reasonable and does not endanger the purpose of this investigation as there is consistency within such a field between simple and more involved objects.

These fields of science are:

1. classical thermodynamics [2],
2. statistical mechanics [3],
3. information theory [4; 5; 6].

The object to be considered will be

one mole of a perfect gas.

* Completed and revised version of a lecture given at Birlinghoven at the "Gesellschaft für Mathematik und Datenverarbeitung", 15. 8. 1972.

2. The object to be considered

This object "perfect gas" has particular aspects in each of these three fields and is thereby connected to that version of entropy which refers to it.

A perfect gas is an artificial creation of the human mind which does not exist in nature when applying strict criteria. Nevertheless this notion has a meaning in the real world since any real gas approaches asymptotically to the properties of this perfect gas when diluted. This corresponds to the physical definition of a perfect gas, that its total internal energy does only consist of the kinetic energy of its microphysical particles. Per degree of freedom, the kinetic energy is proportioned to temperature. To simplify the considerations, each molecule of the gas is assumed to be monatomic and to have only three translatory degrees of freedom. (There are no rotatory and no internal degrees of freedom in such a molecule).

One mole of gas is an aggregation of N such molecules where $N = 6{,}02 \,.\, 10^{23}$ molecules/mole. On this occasion, we may add that the kinetic energy per molecule with three degrees of freedom is

$$\frac{U}{N} = \frac{3}{2} kT, \tag{1}$$

where U is the internal energy of one mole of gas, T the absolute temperature and $k = 1{,}38 \,.\, 10^{23}$ Joule/centigrade Boltzmann's constant. (1 Joule = 1 watt.sec.)

3. The object of classical thermodynamics

Macrophysically, the state of a perfect gas is indicated by the values of its *macrophysical state quantities* which are its volume V, its temperature T and its pressure p. Experiments have confirmed the *empirical law* that these three state quantities, of a perfect gas, depend on each other according to the formula

$$pV = RT \tag{2}$$

where there is the universal constant

$$R = kN = 8{,}27 \text{ Joule/centigrade . mole.} \tag{3}$$

Another *empirical law* is the first principal theorem of thermodynamics which reads that within a *closed* system the internal energy

$$U = Q + \bar{Q} \tag{4}$$

is a constant, Q being heat and $\bar{Q}$ non-heat (= the sum of all energies with exception of heat). Furthermore, within a closed system, there is an empirical fact that in every process that really occurs, non-heat is converted into heat. Expressing this in other words renders the *empirical law*, that there is an asymmetry between macrophysics and microphysics. The energy exerted by a macrophysical body is liable to be split up into kinetic energy of the individual particles. This dispersion is irreversible.

Facts and relations between facts are given by experience. Mathematical inference is to expand the applicability. By substituting equation (3) into equation (1) we get

$$U = \frac{3}{2} RT, \tag{5}$$

which is the macrophysical correspondence to equation (1). This equation is not to be construed in such a way that the gas contains only heat; this will only hold after all kinetic energy is completely used as heat. Since this gas exerts a pressure upon the enclosing walls, mechanical energy can be drawn away by expanding it against an external force. A differential increase of the volume by dV against the pressure p will produce the work

$$-dA = p\,dV \tag{6}$$

on account of the internal energy. (The minus-sign on the left side of this equation is to indicate that this energy *leaves* the system when the volume is increased). Differentiation of equation (5) gives

$$dU = c_v\,dT; \qquad c_v = \frac{3}{2} R. \tag{7}$$

After differentiation of equation (4) and substituting there $\delta \bar{Q}$ by dA which is the only non-heat in this system we will have

$$\delta Q = p\,dV + c_v\,dT, \tag{8}$$

or, after substituting for p from equation (2):

$$\delta Q = R\left(T\frac{dV}{V} + \frac{3}{2}dT\right). \tag{9}$$

Due to the asymmetry mentioned previously, *this quantity can never be negative.* But, although it expresses this quantity in terms of state quantities, the equation (9) fails to define a state quantity due to mathematical implications. The integral on the right side of equation (9) depends of the path of integration chosen in the $(V - T)$-plane. Therefore, a value Q calculated in this manner not only depends on the present state quantities but on their historical past. Put in mathematical terms, the right side of equation (9) is not a total differential. Obviously, this difficulty can be mended by dividing this equation by T. On the left side, we may introduce a new differential $dS = \delta Q/T$. *Since T is always a positive real quantity,* dS *will remain non-negative as long as* δQ *is non-negative.*

Integration of the new equation

$$dS = R\left(\frac{dV}{V} + \frac{3}{2}\frac{dT}{T}\right) \tag{10}$$

results in

$$S = R\left(\ln V + \frac{3}{2}\ln T\right) + S_0. \tag{11}$$

This state quantity which is a measure of irreversibility is called *entropy*. (The constant S_0 is undetermined for the time being. It will be considered in section 7.) To give some physical elucidations about the "meaning" of entropy, we might begin with the statement that entropy is a state quantity like the internal energy U. Its use is to lead to a neat formulation of the second principal theorem of thermodynamics just as U does in respect to the first principal theorem. Entropy is not more abstract than is energy where we are accustomed to this abstractness.

Physically, by returning to the very origin of this expression may help to get a grasp on it. By "integration" of the (mathematically senseless) formula

$$S = \int \frac{\delta Q}{T} \tag{12}$$

we get entropy. Thereby, this formula may be considered as a symbolic instruction how to get entropy: At each differential element of time and in any element of volume, there may be a transfer of no-heat into heat. Every such elementary transfer has to be divided by the temperature of this element before summing up. The sum is entropy.

We may compare this situation with the transfer of loss of material during a production process into costs. Since the material increases its monetary value during processing (e.g. iron is gradually transferred into watch-springs), summing up the losses of material and then considering this loss as another expression for costs is not correct. Instead, we should transfer each individual elementary loss into costs and then sum up.

4. Possible thermodynamic processes within a perfect gas

We consider a closed system containing one mole of a perfect gas and a device having arbitrary properties within the range of valid physical laws. The question is: Which is the boundary line enclosing the field of possible states and processes within this system when an initial state is given? Thereby, the device can exert all constraints not contravening the first and the second pricipal theorem of thermodynamics. Explicitly this means (Fig. 1):

1. Assuming that the device does not supply energy, there is a limit to temperature according to equation (5) which is independent of the actual volume of this gas. This limit is a horizontal. No state can be above this line.

2. A line connecting the points of equal entropies is called an isentrope. In a plane where the logarithms of the two variables are plotted down, these isentropes are straight lines. No state can lie on an isentrope whose entropy is lower than that of the initial state.

3. Since $\mathrm{d}S$ can never become negative, no path within the $(V - T)$-plane is permissible cutting an isentrope in reverse direction (i.e. direction of falling entropy).

The limit case is given when a path coincides with an isentrope. Such a process is only possible in the imagination. This process is reversible.

A reversible process which is impossible due to the impossibility to meet the perfect technologigal requierments is not to be confused with a reversible process which is even then impossible in the mind when perfect conditions are assumed.

These limits to realization may be demonstrated, in two typical experiments:

Experiment 1: Expansion of a perfect gas into vacuum (Fig. 2).

In a closed system where no non-heat is stored, the equation (1) connects a constant internal energy U to a constant temperature T. Therefore, the second term will vanish at equation (11). As the volume increases to twice its initial value, the entropy increases by $\Delta S = R \ln 2$. This irreversible process is represented by a path along the horizontal boundary until the volume limit is attained.

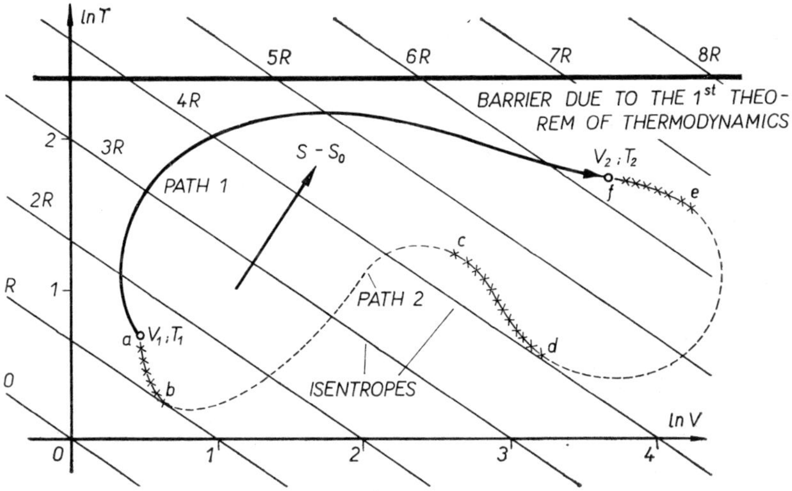

Fig. 1. *Possible and impossible processes.* This plane represents the thermodynamic states of a perfect gas. A path connecting a succession of points represents a process. Since ln V and ln T are plotted down, the isentropes are straight lines. Path *1* is possible, path *2* is not. (×-×-×-× indicates sections of the path contravening the 2nd theorem of thermodynamics.)

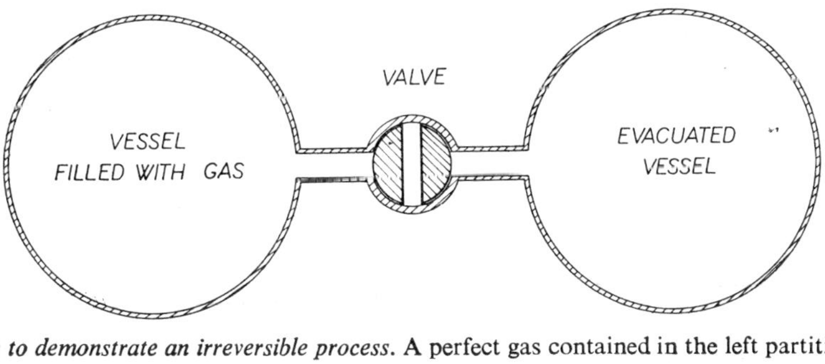

Fig. 2. *A device to demonstrate an irreversible process.* A perfect gas contained in the left partition of the vessel will expand into the evacuated right section after the valve is opened.

In this process there is, at the beginning, a pressure difference which means a store of work. After opening the valve the gas is accelerated thereby assuming macrophysical kinetic energy. Due to turbulence within the gas and friction between the gas and the including walls, this macrophysical kinetic energy is dissipated into microphysical kinetic energy which is heat.

Experiment 2: Expansion of a perfect gas whereby the work at disposal is removed and stored (Fig. 3).

During the expansion, the work at disposal is drawn away and stored into an imaginary perfect device. Since there is no non-heat left to be transformed into heat, the entropy does not increase and the process is forced to follow an isentrope. By using the stored work, this process can be reverted at any moment.

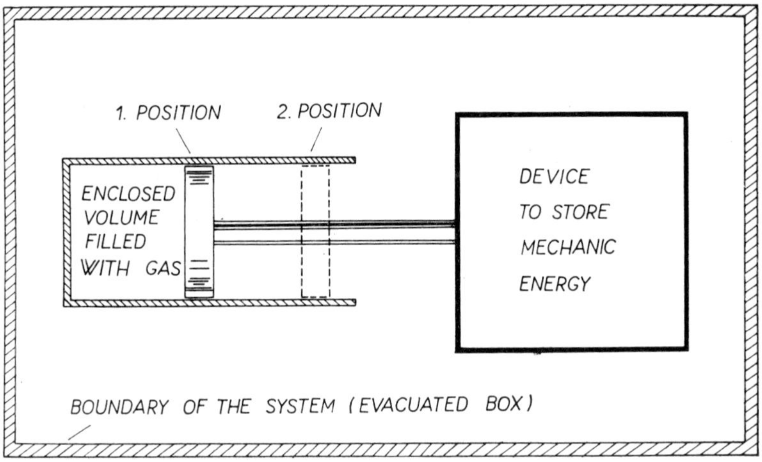

Fig. 3. *A device to demonstrate a reversible process.* A perfect gas is slowly expanding against an external force thereby delivering the work set free into a store. In reverse direction, the gas is slowly compressed whereby the work which is necessary is supplied from store. In perfect conditions, this process is reversible.

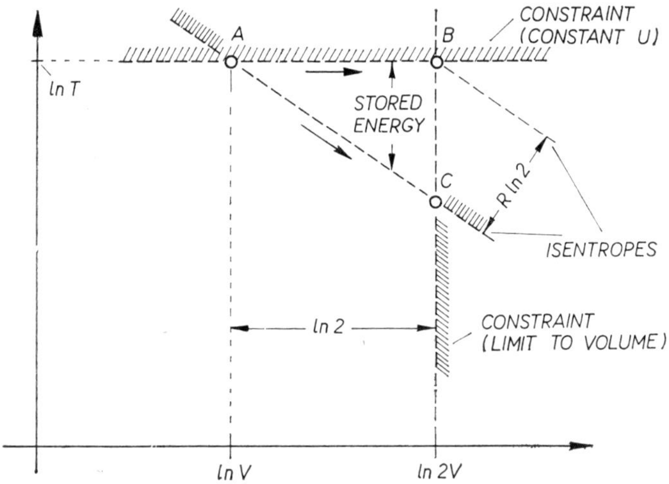

Fig. 4. *Comparison of the two processes in the state plane.* When the volume is increased to $2V$, then the path will proceed from A to B. But if, during the expansion process, the work set free is continuously removed, then the path will proceed along an isentrope from A to C.

Comparison of these two processes

On the logarithmic state plane, both processes may start at point A. This point (Fig. 4). represents the highest possible entropy under the prevailing constraints. In experiment 1, the volume constraint is altered to a higher value. The state follows until it becomes stationary again at the highest possible entropy. In Experiment 2, only a path along an isentrope is possible. Therefore, the state can only move to C.

5. Entering into the scope of entropy

Generally, any process going on in a closed system increases the entropy to the maximum value within the existing constraints. This is due to microphysical processes where each of them is irreversible as long as there is still a margin to the maximum of the possible entropy.

A measure of irreversibility may appear to be a strange expression as it is difficult to comprehend a measure of an event which is impossible. But thus impossibility refers only to the closed system. By brute force from outside the system, a reversal may be possible, e.g. the expansion of the gas in Experiment 1 can be reverted with help of an external air-pump. In perfect conditions, this process can be reversible. When comprising these two systems into one closed system, the entropy reduction of the "irreversible" system is equal to the increase of entropy in the attached external system. This entropy is a measure of irreversibility. In this context, it can be called the measure of stability, too.

6. The entropy of mixing

According to general experience in every-day life, virtually all things that lose their present state do so irreversibly. Considered as a general measure of irreversibility, the total entropy in the world, therefore, is to increase monotonously. Using Experiment 1 and 2 as typical examples, there is perhaps one difference: Some processes are irreversible due to fundamental reasons that cannot be overcome by perfect conditions (Experiment 1); others are irreversible in reality as the perfect conditions cannot be realized by any available technology. This applies to the so-called reversible processes (Experiment 2). Mixing belongs to one of these two types. The ruling doctrine considers mixing to be irreversible for fundamental reasons; this paper prefers to associate this type of irreversibility to technological imperfections.

If mixing is intrinsically connected to an increase of entropy even under perfect conditions, then separation can only happen by interaction from another system. The answer to this question can only be found with help of an imaginable perfect process. A crucial one was found by Max Planck towards the end of last century and which we call the

M. Planck's Experiment (Fig. 5).

M. Planck uses two diaphragms [2, p. 213] having complementary properties. One of them is permeable for gas A_1 and impermeable for gas A_2; the other has the same properties with the gases interchanged. When such a diaphragm is used as a separating wall between mixtures of gases, the permeating gas will flow as long as there is still a difference in the *partial* pressure of this gas at both sides of the wall. This is independent of the partial pressures of the rejected gases. After the flow has ceased due to a balance of the partial pressures, then, at each side,

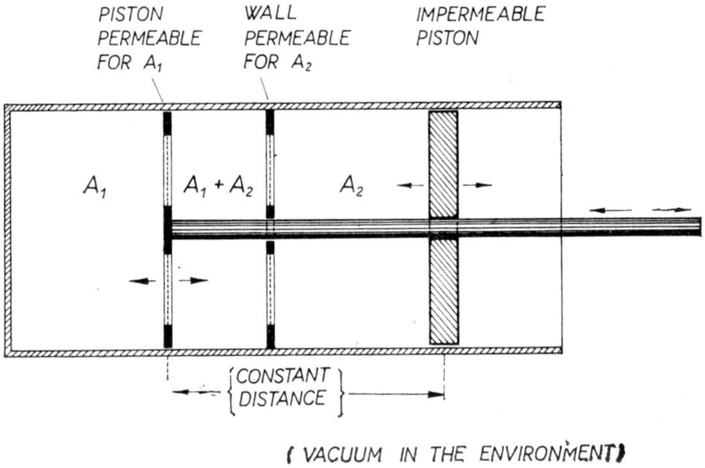

Fig. 5. *The model of M. Planck.* The piston and the wall which include the mixture are permeable for one of the gases, the piston (say) for A_1 and the wall for A_2. The permeable piston is connected to an impermeable piston at a constant distance. At one extreme position of the double piston, both gases are mixed; at the other, they are separated, whereby each volume is equal to that of the mixture.

the rejected components of the gas mixture exert their partial pressures on the wall. In Planck's device, two pistons are attached to a common rod which makes the forces balance at any position of the double piston after the flow has ceased. This process is therefore reversible which means that the total entropy does not change. After an intelligible generalization of more than two gases, Planck is able to confirm a statement by Gibbs. To quote Planck:

Statement I:
"*The entropy of a mixture of gases equals the sum of the entropies of the individual gases, if each of these alone occupies the whole volume of the mixture at the same temperature.*"

Perhaps there may be doubt about the physical permissibility of complementary diaphragms. In fact, Planck had made experiments with diaphragms himself but he did not have a complementary pair. (For comparison, there are no complementary macrophysical sieves.) In the decades of time between, a complete theory of diaphragms [7] is established and confirmed experimentally [8].

In the present context, the physical or technological existence of a pair of complementary diaphragms is not the essential point. We may consider these diaphragms to represent symbolically any perfect device that has the faculty to discriminate an existing difference and to separate the objects accordingly.

Having this statement as the background, the existence of mixing entropy can be proved as follows: Let us consider two vessels having the volumes V_1 and V_2 respectively. These are filled with gases A_1 and A_2 at equal pressure and equal temperature. Therefore, the proportion

$$m_1 : m_2 = V_1 + V_2 \tag{13}$$

holds where m_1 and m_2 are the numbers of moles of the two gases. After the vessels are connected together (e.g. after opening a valve), the gases will diffuse into each other due to thermal agitation. In the special case $V_1 = V_2$, according to Gibbs, entropy is preserved when the mixture has the volume $V_{12} = V_1 = V_2$. Mixing, therefore, is an expansion to a higher volume, in so far as entropy is concerned. Generally, this increase is equal to the expansion entropy according to Experiment 1 of the two gases independent of each other. The total sum of this increase in entropy is

$$\Delta S = R\left(m_1 \ln \frac{V_1 + V_2}{V_1} + m_2 \ln \frac{V_1 + V_2}{V_2}\right). \tag{14}$$

By using the abbreviations $z = (V_1 + V_2)/V_1$ and $m = m_1 + m_2$, this expression can be simplified to

$$\Delta S = mR\left(\ln z - \frac{z-1}{z} \ln (z-1)\right). \tag{15}$$

This covers a pure expansion process as in Experiment 1 with the special values $m = 1$; $z = 2$.

The presuppositions are empirical facts, this derivation is correct (and sanctified by its age). But a controversy is left which should be explained. At the present moment, arguments are directed against the equivalence between mixing and the superimposition of expansion processes:

1. As long as only thermodynamic state properties are considered, which are pressure and temperature in macrophysics and position and momentum in microphysics, there are no different properties between two perfect gases. There *are* no different properties between two perfect gases in thermodynamics; all gases must behave equally and equal behaviour will lead to equal state properties, one of which is entropy.

2. In looking at equation (14) this entropy of mixing is completely equivalent to two expansion processes into *vacuum*. From the point of view of energy transfer, this equivalence cannot hold. At the expansion into vacuum, there is real potential mechanical energy set free and subsequently and gradually transformed into heat. In a mixing process, nothing of this kind occurs.

3. Gibbs has stated a paradox by saying that if one gas is gradually modifying its property so that it continually approaches to that of the other one, then at last a jump in entropy must occur, which is absurd. In the accepted doctrine, this problem was settled by raising the objection, which is correct, that two gases are either different or identical; there is no continuous transition from one property to the other. But then an isotopic difference is sufficient to cause a difference.

4. There is, however, a continuous transition from mixing macrophysical particles (e.g. peas and beans) to the mixing of the particles of a gas. Macrophysical particles can be mixed and separated again by perfect transport systems. At what state of this transition does entropy of mixing come into play?

Certainly these arguments are not strong enough besides creating an unpleasant feeling which sometimes is a sufficient incentive to investigate further. At this moment, for the sake of being in the position to resume the discussion later on, as a reference we want to establish tentatively the following

Statement II:

Mixing perfect gases of equal pressure and equal temperature does not increase entropy. After mixing, the gases together will occupy the sum of the volumes that they had individually before at the same pressure and at the same temperature.

This statement of course holds for the trivial case, that different gases are really identical.

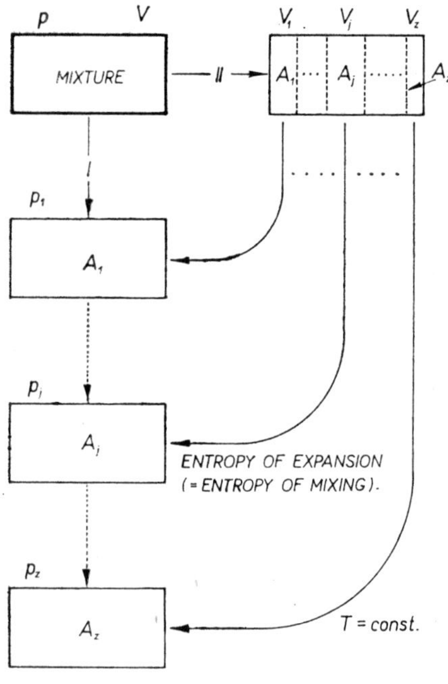

Fig. 6. A mixture of gases is separated in two different ways, in procedure I with the help of the Planck device, in procedure II by any other suitable device. (The entropies necessary for expansion may be supplied from outside.) There is an entropy difference between these two aggregations of separated gases which at constant temperature is equal to the sum of the individual entropies of expansion.

Now we want to derive a general formula for the entropy of mixing by using these two statements. Let us assume a mixture of the gases A_1, A_2, ..., A_j, ..., A_z, which in total is just one mole (Fig. 6.). The share of the individual gas A_j is indicated by its mole-figure m_j. This gas may be dissected into components according to the two modes in accordance with Statements I and II.

In dissection I, each individual gas has got the same volume V of the mixture. Then at constant temperature T, the gas has the pressure $p_j = m_j p$ where p is the pressure of the mixture.

In dissection II, each individual gas has got the same pressure p of the mixture. Then at constant temperature T, the gas has the volume $V_j = m_j V$ where V is the volume of the mixture

In both cases, the total entropy of the mixture is the sum of the entropies of the individual gases when applying the associated statements respectively. As the statement holds when mixing identical gases, the entropy of mixing is the sum of the entropy differences in the individual gases when comparing the entropies of one gas with an other that arise from the two different modes of dissection.

In dissection I, gas A_j has the volume $V_j = m_j V$. Comparing the entropy of the same gas in dissection II where it has the volume V is for m_j moles equivalent to an increase in entropy by

$$m_j R \ln \frac{V}{m_j V} = -R m_j \ln m_j .$$

Therefore, the total increase is the sum

$$\Delta S = -R \sum_{j=1}^{z} m_j \ln m_j . \tag{16}$$

This derivation, of course, presupposes that Statement II only holds for the trivial case that the individual gases are not different. Then, if they are really different, the total entropy increases by the mixing entropy ΔS.

We acknowledge this statement to be justified according to the present state of recognition. After having prepared another platform for arguments, which will happen in the next sections, we want to resume the proceedings in Section 9 of this paper.

7. Transition to microphysical entropy

The purpose of this section is not to introduce the entropy of microphysical particles into discussion but to demonstrate the close connection between these two presentations of entropy. At first glance, when clinging to the notion of irreversibility, there seems to be nothing that is irreversible in microphysics. All laws that command the motions of microphysical particles are reversible. They do not change their validity when the time t is replaced by $-t$. Since there is a danger of delay by an immediate philosophical attack on this problem, we prefer the attitude of heuristic formalism trusting in chance to find an interpretation during this work or afterwards.

In a naive sense, formula (11) represents the entropy of N microphysical particles. Dividing by N gives

$$s = k\left(\ln V + \frac{3}{2}\ln T\right) + s_0 , \tag{17}$$

where $s = S/N$, $s_0 = S_0/N$, $k = R/N$. To consider this to be the microphysical entropy of an individual particle does not make sense since V and T are not coordinates in microphysics.

The microphysical state of a particle is to be described in the coordinates

$$\Omega_i = (x_1, x_2, x_3, y_1, y_2, y_3), \tag{18}$$

where x_1, x_2, x_3 are the components of the position and y_1, y_2, y_3 the components of the momentum. Such a description is, however, impossible. The individual particle under consideration will remain an active member of the family of particles from which it is selected at random. As such, it is performing collision processes in the six-dimensional space Ω. The description of a succession of definite states Ω_i in this space must, therefore, be replaced by the description of the distribution of the probability of this particle to be in the state Ω_i.

In a perfect gas, this probability $p(\Omega_i)$ can be substituted by the product

$$p(\Omega) = p(\boldsymbol{x})\, p(\boldsymbol{y}), \tag{19}$$

where $p(x)$ is the probability to occupy a definite point $\boldsymbol{x}_\mu = (x_1, x_2, x_3)_\mu$ and $p(y)$ the probability of a momentum vector $y_\nu = (y_1, y_2, y_3)_\nu$ in the momentum space. This substitution is permissible due to the fact that the position of a particle is independent of its momentum and vice versa. There are individual constraints for the position and for the momentum at this special case. Since formula (17) contains two additive terms whereas expression (19) is a product, the heuristic idea may appear that the microphysical counterpart to expression (17) might be $s' = K \ln p(\Omega)$, which can be split up into $s' = K \ln p(\boldsymbol{x}) + K \ln p(\boldsymbol{y})$. Now, a comparison term by term appears to be possible where the constant K is to be determined.

Since a function of a probability has a fluctuating value which is not comparable to a determined value, there is reason to use the expectation value

$$s = K \sum_i p(\Omega_i) \ln p(\Omega_i). \tag{20}$$

By splitting up this expression into two terms and by equating to equation (17) term by term, the two equations arise:

1. $$K \sum_\mu p(\boldsymbol{x}_\mu) \ln p(\boldsymbol{x}_\mu) = k \ln V + \alpha_1, \tag{21a}$$

2. $$K \sum_\nu p(\boldsymbol{y}_\nu) \ln p(\boldsymbol{y}_\nu) = k \frac{3}{2} \ln T + \alpha_2. \tag{21b}$$

where $\alpha_1 + \alpha_2 = s_0$.

In these two equations which are to prove that both sides of them, when added, represent numerically the same entropy, the probability functions $p(\boldsymbol{x})$ and $p(\boldsymbol{y})$ have to be determined first. In this special case of independence, these are the functions where the left sides assume individually their maximum values regarding the constraints.

These constraints are

1. for the left side of equation (21a):

the probability $p(\boldsymbol{x})$ is zero for an $\boldsymbol{x}$ outside the volume V which is enclosed by the walls of the container,

2. for the left side of equation (21b):

the total kinetic energy of all particles is N times the expectation value

$$\frac{1}{2m}\sum_{\nu} y_\nu^2 p(y_\nu) = \frac{3}{2} kT, \tag{22}$$

where m is the mass of one particle.

Moreover, we may refer the left sides to a perfect measuring device where the power of resolution is represented by the volume element V_0 and the momentum element y_0. These two values are dependent on each other according to the Heisenberg uncertainty law, which gives

$$V_0 y_0 = \left(\frac{h}{2\pi}\right)^3. \tag{23}$$

In this equation, h is the constant 6.62 10^{-34} Joule . sec of Planck.

Now the problem is completely determined. The element V_0 may have an arbitrary value. Since this determines y_0 by equation (23), we will see later on, that this value does not influence the final result.

In equations (21a) and (21b), the functions $p(x)$ and $p(y)$ have now to be determined. This requires a general principle concerning the dependence of this function upon the constraints. Statistical mechanics shows that the collision processes of the microphysical particles change the distribution function of probability continuously until $-\sum_i p(u_i) \ln p(u_i)$ has attained its maximum value at the prevailing constraints. Then the system is in a statistical equilibrium. (Here, u is the random variable). The determination of $p(u)$ is a straightforward problem of the calculus of variations. This principle moreover, is extremely interesting in the general context since it introduces again irreversibility also into microphysical entropy.

In the case of constraint 1 for equation (21a), the result is a very simple and intelligible one, supplying $p(x_\mu) = V_0/V$ independent of μ. There is an equipartition of probability within the volume which is occupied by the gas. In the case of equipartition, equation (21a) will reduce to

$$K \ln p(x) = k \ln V + \alpha_1. \tag{24}$$

Since $\ln p(x) = \ln V_0 - \ln V$, equation (21a) is verified with the values

$$K = -k; \qquad \alpha_1 = -k \ln V_0. \tag{25}$$

The same procedure has to be used to determine $p(y)$. Some minor alterations may be applied to simplify the mathematical treatment:

In the momentum space, all three components are equivalent and the probability in reference to one component is independent of that in reference to another component. Therefore, equation (21b) holds for one component alone when divided by 3. Furthermore, the sum can be substituted by an integral. On this occasion, the resolving power of the perfect observer, which is represented by y_0, is to be considered.

Taking together these implementations, we introduce the density of probability (in reference to onc component)

$$P(y) = \frac{p(y)}{3y_0}, \tag{26}$$

which will permit to transform equation (22) into

$$Y = \int_{-\infty}^{+\infty} y^2 P(y)\,\mathrm{d}y = \frac{1}{2} \,.\, 2m \,.\, \frac{3}{2} kT = mkT. \tag{27}$$

The solution of the variation problem, which now is the probability density of *one* component of the momentum, renders

$$P(y) = \frac{1}{\sqrt{(2\pi Y)}} \mathrm{e}^{-y^2/2Y}. \tag{28}$$

Instead of equation (21b), we now have

$$K \int_{-\infty}^{+\infty} P(y) \ln P(y)\,\mathrm{d}y + \frac{1}{3} K \ln y_0 = \frac{1}{2} k \ln T + \frac{1}{3} \alpha_2. \tag{29}$$

We substitute for this equation

$$\int_{-\infty}^{+\infty} P(y) \ln P(y)\,\mathrm{d}y = -\frac{1}{2}(\ln T + \ln 2\pi mk + 1),$$

$$\ln y_0 = 3 \ln \frac{h}{2\pi} - \ln V$$

and the identifications in equations (25).
This verifies this equation too with

$$\alpha_2 = k\left(\frac{5}{2} \ln 2\pi + \frac{3}{2} \ln mk + \frac{3}{2} - \ln h + \ln V_0\right).$$

By this comparison between macrophysical and microphysical entropy, the constant s_0 now, which was introduced in equation (17), will receive a definite value

$$s_0 = \alpha_1 + \alpha_2 = k \ln\left(\frac{2\pi}{h} (2\pi\, emk)^{3/2}\right), \tag{30}$$

where $\ln V_0$ cancels out according to anticipation. In this equation, k is Boltzmann's constant, h Planck's constant, m the mass of the particle and e the basis of natural logarithms.

By this successful identification, evidence is given that the macrophysical entropy according to equation (17) and the microphysical entropy

$$s = -k \sum_i p(\Omega_i) \ln p(\Omega_i) \tag{31}$$

are intrinsically identical. Since now the constant s_0 has a definite value, the entropy has a definite value, too. Only this constant depends of the mass of the particle. Also in microphysics, entropy is a measure of irreversibility (considered from inside the system) and of stability (against interaction from outside).

Regarding the mathematical form which is called Boltzmann's entropy, it is an evaluation of the spread of the probability distribution function. The wider the spread the higher are irreversibility and stability.

For an observer who is looking for a precise determination of the microphysical state, entropy means uncertainty. Due to fundamental properties of statistical mechanics, entropy tends against its maximum value which is possible at the prevailing constraints; this is highest spread of the probability function and highest uncertainty in respect to an individual microphysical state, but highest irreversibility and stability as concerns the microphysical probability distribution and the macrophysical state.

8. Information theory contributions

Information theory is pure mathematics since it does not originate from empirical facts but from abstract definitions alone. Therefore, the statements of information theory apply everywhere in handling mathematical expressions of the same kind. Conversely, there is no legitimate possibility to apply formulas that indicate physical relations in an arbitrary mathematical context unless these formulas actually are pure mathematical relations (which means that they do not depend implicitly on any empirical context).

The restricted definition of information relates to a set (a) of symbols where each individual symbol a_i has a probability $p(a_i)$. Then, there is associated to this symbol the information

$$I(a_i) = -\operatorname{ld} p(a_i), \tag{32}$$

where ld designates the logarithm with the basis 2.

The set of symbols is evaluated by the expectation value of this information

$$H(a) = -\sum_i p(a_i) \operatorname{ld} p(a_i) \tag{33}$$

which does not depend on anything else but the probability distribution. C. E. Shannon has given this expression the name entropy, too, in regard to the mathematical form which again expresses uncertainty.

To have a superior expression comprising entropy of thermodynamics and entropy of information theory we want to use in this paper the expression **spread,** which is short for: measure of spread of the probability distribution function. Only when this spread refers to the distribution function of microphysical state properties will it be called entropy, when it refers to a quantity or a set of symbols not representing thermodynamical state properties it will be called **in**formation **t**heoretical ent**ropy**, short: **intropy.** This does not necessarily imply that these symbols are used for communication.

Now we want to generalize the notion information. Information is created by an act, by an event which reduces the a priori spread of any quantity. The numerical difference (a priori spread − a posteriori spread) is called information. In a limit case, the spread is totally removed. This happens when an individual symbol is selected out of a set of a priori possible symbols. If anywhere in a system such a reduction of spread takes place, the spread is also reduced in other quantities which are statistically dependent on it. What we require now is a mathematics of spread which is supplied by information theory. The conspicuous achievement of information theory is to have established and used a very general mathematical model of communication: A transmitter disposes of a set of symbols (a); a receiver is sensitive against set of symbols (b). There is communication if the symbols (b) are statistically dependent on (a). This property to be statistically dependent is represented by a channel.

Due to its general validity, this model can be applied to any statistical dependence of one set of chance values on another set of chance values. The statistical dependence is characterized by a matrix of conditional probabilities $p(b \mid a)$, where the element $p(b_j \mid a_i)$ of this matrix is the probability that b_j appears at the output of the channel when a_i is applied to its input. This matrix $p(a \mid b)$ is an abstract and a very general representation of the channel whereas $p(a)$ represents the transmitter.

There are some relations in probability theory that can be used now: The probability to have at the same time a_i at the input and b_j at the output is

$$p(a_i; b_j) = p(a_i)\, p(b_j \mid a_i). \tag{34}$$

The probability to have b_j at the output without presupposing any knowledge about a_i is

$$p(b_j) = \sum_j p(a_i; b_j) = \sum_j p(a_i)\, p(b_j \mid a_i). \tag{35}$$

The change from $p(b_j)$ to $p(b_j \mid a_i)$ changes the information at the channel from $-\mathrm{ld}\, p(b_j)$ to $-\mathrm{ld}\, p(b_j \mid a_i)$. It is increased by the transinformation

$$\begin{aligned} I(a_i; b_j) &= -\mathrm{ld}\, p(b_j \mid a_i) + \mathrm{ld}\, p(b_j) = -\mathrm{ld}\, \frac{p(b_j \mid a_i)}{p(b_j)} = \\ &= -\mathrm{ld}\, \frac{p(b_j \mid a_i)\, p(a_i)}{p(b_j)\, p(a_i)} = -\mathrm{ld}\, \frac{p(a_i; b_j)}{p(a_i)\, p(b_j)}. \end{aligned} \tag{36}$$

When j is running through all values we get the set of all transinformations which are caused by an individual a_i at the input. Its expectation value is the spread at the output

$$R(b \mid a_i) = -\sum_j p(b_j)\, \mathrm{ld}\, \frac{p(a_i; b_j)}{p(a_i)\, p(b_j)}, \tag{37}$$

in reference to an individual signal at the input. By referring to the complete set of input signals, the transentropy is

$$R(a; b) = -\sum_a \sum_b p(a; b)\, \mathrm{ld}\, \frac{p(a; b)}{p(a)\, p(b)}. \tag{38}$$

Transentropy surprises by being a symmetric property between two points of a system. The average information that an observer at the output receives in respect to the input symbols equals the information that an observer of the input signals can get in respect to the actually received output signals.

This transentropy is not a property of the channel alone; it depends on $p(a)$. This is evident. When the probability distribution prefers those signals at the input which have a high transinformation than this must increase the expectation value. In fact, there is a general law [5, p. 138; 6, p. 181], saying that this distribution must be chosen so that each individual symbol is contributing to the transentropy with an equal share. That quantity which is gained by maximizing the transentropy with respect to $p(a)$ is called the channel capacity

$$C = \max_{p(a)} R(a;b). \tag{39}$$

By quoting the principal theorem of information theory which was established in 1948 by Shannon we may say in other words:

In general, the intropy reduction at the output of a channel is lower than that at the input. Only in extreme cases (which are attained when the generator of the input spread is optimum matched to the system by choosing an appropriate probability distribution), *the output reduction approaches to the input reduction.*

To emphasize we may repeat that this is a pure mathematical theorem where nothing about the physical (or non-physical) character of the spread is presupposed. In no case can the reduction of intropy at the output be higher than that reduction at the input.

Since the spreads are mutually dependent we are not permitted to discriminate between input and output. Referring to a *closed* system, the theorem therefore holds:

Within a closed system, the sum of all spreads of all quantities (*not only thermodynamic ones*) *can never decrease. In a perfect case, the spread of one quantity can be exchanged against the spread of another one.*

9. Resuming the entropy of mixing

After this intermediate treatise, the mixing problem may be resumed. We want to reconsider the z boxes filled with different gases, where the individual box j is filled with m_j moles of the gas A_j. All boxes have an equal volume V. After mixing, a box of equal volume as the individual ones will contain one mole of mixture. Now let us use a perfect device which can distinguish between the different gases. One particle selected at random from the mixture is a member of the subset A_j of all particles with the probability m_j. The detection of the membership produces the information $-\mathrm{ld}\, m_j$. Before detection, the members of the subset A_j whose number is $m_j N$, have the intropy $-m_j R \ln m_j$. The total intropy of all particles together is

$$\bar{S} = -R \sum_j m_j \ln m_j. \tag{40}$$

This is numerically the same quantity as in formula (16). The only difference is that now the m_j are probabilities of the individual particles being a member of a definite subset.

This permits one to replace the Gibbs theorem by another one by referring to the experiment of M. Planck which now will receive a different interpretation:

1. *If the designation "entropy" is used only as a measure of spread of the probability distribution function of the microphysical thermodynamical state whereas the designation "intropy" is used as an analoguously defined measure of another quantity then the sum entropy + intropy can never decrease within a closed system.*

2. *In perfect conditions, entropy can be exchanged reversibly with intropy.*

In Planck's experiment, this non-thermodynamic property is the membership of an individual particle to a subset A_j of all particles. In a mixture of gases, this spread is measured by $\bar{S}$ according to formula (40). When the gases are separated, then each particle is confined to a box labelled with the appropriate membership. There is no spread, no intropy. When defining information as the result of reduced intropy [9, p. 14 and 18]; then information appears in the separated gases but is destroyed when mixing them.

There is no difference between the mixing processes of different gases when compared with that of identical gases when the device used is not sensitive to the discriminating properties. If such properties are present and if the device is sensitive to these properties, then the increase of spread of these properties (intropy) can be used to reduce entropy.

Therefore, the notion "entropy" of the second principal theorem of thermodynamics must either be replaced by spread (= entropy + intropy) or be redefined to include intropy. This is not advisable since this notion would then become a property which depends not only on the object but also upon the properties of the measuring instrument. Conversely, when keeping to the propositions of this paper, the actual thermodynamic property "entropy" will become exchangeable with the spread of other properties *if suitable devices are available.*

10. Physics and information

Finally, in this last stage of consideration, we will use once more the perfect gas as a monitor. In formula (17), with the exception of the constant s_0 which depends on mass according to equation (30), and in formula (31), the entropy does not depend on the size and mass of the particles. As long as there is no question of splitting up a particle into separate particles, of gluing together two or more particles in to one, of destroying and creating particles, we may replace the perfect gas by a macrophysical model (of it). Within a imperturbable perfectly rigid box which is evacuated there are some perfectly elastic balls performing collision processes. The movements are not be influenced by any fields. Again, each ball has three degrees of translatory movements and no internal degrees of freedom. There may be some different classes of balls, e.g. discernible by their colours white, black, green, yellow, red and blue.

Any configuration of these balls within the box which is distinguisted to be one from another is a signal in the sense of information theory. This is therefore comparable with the positions of the particles of a gas. It is *one component* of thermodynamic entropy. The second component which has to be considered simultaneously, is the momentum of an individual particle, the total kinetic energy, the temperature. The colour now is an additional property which is transferable into reduction of entropy if suitable devices are present.

Such an aggregation of balls in comparable to the model of Planck. By a reversible process within this system, a peculiar pattern can be selected which is the destruction of intropy by simultaneously creating entropy. Conversely, a peculiar pattern can be destroyed which is increasing intropy at the expense of entropy. Let us call such a system an entropy/intropy converter or for short a converter.

Conversion is only then reversible if there is even no exchange of information with the environment. A converter which produces information can be called a transmitter or in another application an actuator. A converter which receives information is a receiver or a sensor. This information is carried by a signal from transmitter to receiver. A receiver is only then sensitive to a signal if this delivers the non-heat energy [6, p. 235]

$$\bar{Q} = kT \ln 2H \tag{41}$$

to the input of the receiver. This is to be calculated by equating $\Delta Q/T = k \ln 2H$, where $\Delta Q = \bar{Q}H$ is the information being received. (Reception means an increase of information which is stored in the receiver system.)

This, however, is only the minimum energy if $kT \gg hf$, where f is the frequency of the signal. A general expression is the formula

$$\bar{Q} = \varepsilon \ln 2H \tag{42}$$

where ε is the energy of Planck's oscillator

$$\varepsilon = \frac{hf}{e^{hf/kT} - 1} + \frac{hf}{2}, \tag{43}$$

which was derived by A. Einstein.

In practice, since there is only a small overlapping, the two approximations

$$\varepsilon \to \frac{hf}{2} \quad \text{for} \quad hf \gg kT,$$

$$\varepsilon \to kT \quad \text{for} \quad hf \ll kT \tag{44}$$

are sufficient.

In the transmitter, this energy has to be replaced by energy delivered to the system from outside. Correspondingly a receiver will convey this energy to any other system which is attached to its output. A difficult question arises when we consider a receiver which is a closed system in all other respects (with the exception of receiving information and energy). In this case, information and energy are stored by building up structures and patterns. Conversely, a system which contains stored information can reduce entropy. If for instance an owner-builder has a plan for his house and has made the necessary applications to the authorities and has made arrangements with the architect, the builders and also with his bank, he (in theory) has only to press the button. The information contained in all these arrangements suffices to remove disorder by erecting a structure. The most impressive examples of such converters which reduce entropy are given by nature.

We are non at the end of our consideration which started with irreversibility in macrophysics. After transformation to microphysical state coordinates and by fitting information theory into it, we have now advanced to a rather strong type of information which is firmly attached to matter and energy. There is no necessity to reject any considerations which lead beyond the scope of this paper, but within scope, there is sufficient reason to consider information as a part of this real world.

11. Summing up what entropy appears to be

To investigate the meaning of entropy, the perfect gas is used as a test object to be run through three essential fields of science where this notion is used: classical thermodynamics, statistical mechanics and information theory. As a general result, entropy is revealing itself as a very comprehensive measure of irreversibility, as seen from outside a system, and of stability, as seen from inside.

In this respect, there is no difference between entropy in macrophysics and entropy in microphysics. One is the transformation of the other to other state coordinates.

Due to the fact that collision processes in microphysics continuously modify the distribution of the probability of a particle to be in a particular state towards that maximum which is possible at the dominating constraints, entropy in microphysics is a measure of the spread of this distribution.

Collision processes depend only on position and momentum of the individual pairs of colliding particles. Therefore, any other quality can influence neither this process nor a quantity depending on this process which is entropy.

A controversy in respect to the ruling doctrine crops up where the mixing of gases increases entropy if the gases are different in any respect. A property which cannot influence a process cannot influence the result either.

Now the mathematical apparatus of information theory is introduced which is applicable to any statistical dependence of two or more fluctuating quantities on each other. If there is such a dependence – which in communication is represented by a channel – the principal theorem of information theory holds. This means: If there is another statistical quantity where the spread is defined in analogy to entropy and if there is an actual dependence by any device then there is an exchange of entropy against the spread of this additional quantity.

Such a device is that model which M. Planck uses to prove the existence of the entropy of mixing. By using selective diaphragms it can distinguish between different gases. This paper proves that the quantity which Planck considers to be the entropy of mixing is identical to the thus defined spread of the probability of a particle to be a member of a subclass of all particles. Now, the Planck device appears to be a reversible converter between thermodynamic entropy and information theoretical entropy.

This recognition requires to modify statements:

1. There is no entropy of mixing.
2. The second principal theorem of thermodynamics only holds in systems not containing any device sensitive to non-thermodynamic quantities.
3. In a system containing elements sensitive to additional properties, the notion entropy

is to be replaced by the sum of all spreads defined in analogy to entropy. With this modification, the second principal theorem is preserved.

Finally, the Planck converter is used to demostrate fundamental properties of transmitters and receivers in communication. In particular, in a signal carrying information, an energy higher than a definite threshold value is necessary. This value is a definite limit to the increase of perfomance in communication.

References

[1] Bertrand Russell: Human Knowledge, Its Scope and Limits. 5. impr. George Allen&Unwin, London 1966.

[2] Max Planck: Thermodynamik. 10. impr. de Gruyter, Berlin 1954.

[3] Richard C. Tolman: The Principles of Statistical Mechanics. 7. impr. Oxford University Press, London 1962.

[4] Claude E. Shannon and Warren Weaver: The Mathematical Theory of Communication, 9. impr. The University of Illinois Press, Urbana 1962.

[5] Robert M. Fano: Transmission of Information. M. I. T. Press, New York 1961.

[6] Johannes Peters: Einführung in die allgemeine Informationstheorie. Springer-Verlag, Berlin—Heidelberg—New York 1967.

[7] F. Auerbach, W. Hort: Grenzgebiete der technischen und physikalischen Mechanik, Vol. 7. Verlag Johann Ambrosius Barth, Leipzig 1931, pp. 706—722.

[8] K. Eger: Über die Diffusion binärer Gasgemische in Kapillaren und porösen Stoffen. Mitteilungen aus dem Max-Planck-Institut für Strömungsforschung und der Aerodynamischen Forschungsanstalt Nr. 51, Göttingen 1971.

[9] Leon Brillouin: Scientific Uncertainty, and Information. 2. impr. Academic Press, New York 1964.

A NON-CONSERVED FUNCTION FOR ORGANIZED SYSTEMS

K. G. Denbigh

Introduction

A considerable number of functions or quantities which would be non-conserved could obviously be "invented" from the field of the life sciences. For instance the number which represents the population of any species is non-conserved, but that would be trivial. To be of genuine interest any nonconserved quantity applicable to biology should: (1) have considerable generality, i.e. in the sense of applying to any part of the universe where organisms exist, quite irrespective of what those organisms might be; (2) bear no necessary relationship to entropy, since otherwise it would be no more than a reflection of the Second Law. The present paper is an attempt to base such a quantity on the concept of an organized system.

First it needs to be remarked that there is a tendency in much of the biological literature, a tendency which is mistaken in my view, to use the words "organization" and "order" interchangeably, as if they were synonyms. This has been done for example by such distinguished biologists as Paul Weiss [1] and P. B. Medawar [2]. This is not merely a semantic confusion but has led to serious errors. In particular it has been supposed that because orderliness and entropy *are* related to each other (inversely and logarithmically) any increase in the degree of organization of a biological system must necessarily result in a decrease of its entropy. This is erroneous as will be shown later.

The fact that "organization" and "orderliness" (leaving "order" aside for the moment) are not synonymous can be readily shown by means of a single counter-example. For this purpose compare a living cell with a crystal. This is a comparison which has often been made, (for example by Schrödinger [3]) but without it having been pointed out that one of the two entities in question is *more* orderly than the other but is *less* highly organized. As regards orderliness the crystal clearly has the advantage for it is indeed a very orderly structure, one in which the atoms or molecules repeat themselves at fixed distances. Whereas in the cell there is no good approximation to a reproducible geometrical pattern and even the temporal pattern (an aspect of cell regularity which has been much emphasized by Goodwin [4]) may not be accurately reproducible from one specimen to another. But if the cell is thus less orderly than the crystal, can it be doubted that it is much more highly organized? In saying this I am relying on what I believe to be the normal usage of the term and this will be analysed in greater detail below. Sufficient to say for the moment that an essential aspect of the notion of organization is *complexity* and it is because of this that we can have little doubt in regarding the cell as being the more highly organized of the two entities. It follows that the two attributes cannot possibly be the same since they display a reverse order of ranking.

It would be out of place to deal at length with the meaning of "orderliness" but for the sake of establishing the above distinction I would say that it has two principal and closely related meanings. The first concerns the extent to which any actual specimen of an entity approximates to the ideal or pattern against which it is compared. For example a crystal is more orderly the smaller are the displacements of its component atoms or molecules from the intersection points of an ideal geometrical lattice and the smaller also is the "spread" (to use Guggenheim's term) of its quantised energy states relative to the ground state. The second meaning refers to populations rather than to single entities. For instance if the various buildings in a street were closely similar in height they would be said to be orderly in this respect, even if they were disorderly in others. For any population the mean values of various properties can be established and the orderliness of the population in regard to any one of these properties can then be expressed by some such measure as the smallness of the standard deviation.

The converse of orderliness is disorderliness and it is disorderliness which is related (logarithmically) to entropy. The details of this well-known relationship need not be gone into here, although it should perhaps be added that the term "disorderliness" must be understood in much more than a purely configurational sense as otherwise its connection with entropy results in paradox [5, pp. 54–56].

The term "order" is used in a large variety of ways and has some special connotations in mathematics and logic. In the present context one should speak, I think, of a "state of order" and regard it as being the ideal reference state, laid down and specified according to certain rules or conventions, which is taken as having 100 % orderliness. Its distinguishability from "organization", like that of "orderliness", may be demonstrated by means of an example – for example the comparison of a perfect wall-paper pattern with a great artist's painting. The former has a more highly developed state of order, whilst the letter has a much greater degree of organization.

What is meant by 'organized'?

This term, like the others which have been discussed, has a wide range of application. One can speak about items being spatially organized (e.g. the parts of a painting) but they can also be temporally organized (for example the notes and bars of a piece of music) or again they can be organized in what might be called logical space, as in the instance of the component parts of a scientific theory or mathematical theorem. Furthermore the items themselves may have physical existence or they may be merely conceptual; or again they may be functional entities such as the items "secretary", "accountant" etc. in the structure of business, or they may be processes or routines. The living body is something much more than an organization of physical structures for it is also an organization of chemical processes which take place continuously, as in an open system. From this point of view any skilled performance of the mind and body, such as speech or piano-playing, is an organization of routines and subroutines and sub-subroutines at many different levels ranging from the psychological down to the biochemical.

The problems which arise in seeking a definition of what constitutes an organized system are extremely perplexing and there is no intention here of attempting their solution. It will be sufficient for present purposes to draw attention to *one* of the necessary conditions without in any way supposing that this is also sufficient. This turns on the notion of complexity: an organized system necessarily consists of parts and sub-parts and these are interconnected[1]). This is not to say however that the set A of all known systems which consist of interconnected parts is the same as the set B of all known systems which are organized. The latter is a sub-set of the former: $B \subset A$. For example any collection of points connected by lines belongs to A but not necessarily to B. This is because a further necessary condition for a system being said to be organised depends on the notion of *function*.[2]) One cannot speak about something being organized without at once raising the issue, what is it organized for? Unless this question makes sense – even if it cannot be answered in a particular instance – we are not dealing with an organized system but rather with a member of the remainder of the set A.[3])

1) The significance of complexity in relation to the ability of living things to reproduce has been very clearly brought out by J. von Neumann in his *Theory of Self Reproducing Automata* [6]. Here he shows in the case of automata that there is a critical degree of complicatedness above which self-reproduction can occur.

2) However it may be that the presence of function is not so much one of the conditions of a system being organized as one of its *consequences;* i.e. an organized system has those particular functions which its constituent parts and its interconnections are capable of achieving for it. This suggests the possibility of a *logical* relationship between structure and function. In some instances this may be one of isomorphism, for example where one structure is capable of reproducing another. Thus a gramophone record can reproduce a piece of music and therefore it would appear that the original score, the grooves on the record and the music itself must all bear to each other a relationship of isomorphism. In other instances, where replication is not one of the functions, it seems likely that the relationship is of a one-to-many type. Thus an organized system, according to this view, is an entity such that the structure of its performance can be mapped into its own structure.

3) For the sake of completeness mention should be made at this point of the class of systems which Prigogine has called *dissipative structures* [7; 8]. Examples of these are mixtures in which there is a partial separation of components due to a temperature gradient, or liquids in which hexagonal convection cells are established, etc. Their general characteristic is the maintenance of a state which is displaced from thermodynamic equilibrium due to a continuous energy influx. Prigogine believes that living creatures belong to this class.

As in the text let A be the set of all known systems consisting of interconnected parts and let B be the set of all known organized systems. Further let C be the set of all known dissipative structures. Then $B \subset A$ and $C \subset A$. However B and C are not identical sets, since one can find many examples of organized systems, such as have been quoted in the text, which do not require an energy influx and one can also find many examples of dissipative structures for which the question "What are they organized for?" would not be meaningful. Yet the sets B and C have an important overlap region—an "intersection"—and one has perhaps made at

Consider some examples. When we are concerned with *manmade* systems such as machines or businesses the answer to the question of function is straightforward.[4]) Indeed in the case of a technological device the answer is what is laid down in a *patent*—i.e. a statement of the function of the device (if it is novel) in relation to the already well-understood functions of its various parts.[5]) On the other hand in the instance of organisms themselves what function they have is beyond human knowledge. No doubt one can speak vaguely about survival, but that is all. Nevertheless the question is still a meaningful one and we have no hesitation in regarding organisms as being organized systems—and indeed *par excellence*.

Can the same be said of atoms and molecules? These are orderly systems but can they also be regarded as being organized? Rather than attempting to resolve this issue I would like to suggest that the concepts of orderliness and of organization, which have been shown to be quite distinct at the level of complex entities, may be much less so at the molecular level. There may in fact be a progressive separation of the two concepts. Putting the matter a little differently, "organization" may perhaps be seen as being a sort of *higher level order*—i.e. in the sense of a hierarchy of ascending types of order where only the lowest level order has a relationship with entropy.

Rather than going further into these issues.[6]) I turn now to an aspect of organized systems which leads directly to the point of this article. This is the notion that "organization" may be additive—i.e. in the sense of allowing one to speak of its "amount" or "degree". Woodger [14, pp. 291–306] has pointed out, for instance, that when cell division occurs the two or more cells which are the result of division may either separate or remain as parts of a single organism. In the latter case, as he says, the organism's degree of organization *has increased* since it now has additional parts each of which is characterized by the organization which previously characterized the original cell.

Woodger's point accords well with the normal usage of language. Thus we often speak of one social institution as being "more" or "less" highly organized than another. These speech forms clearly imply an act of comparison and thus give grounds for expecting that "degree of organization" should be measurable; or at the least that it should be possible to achieve a ranking order within a group of organizations of the same kind.

least a little progress in the definition of "life" in realising that the set D of systems which are said to be living are to be found within this intersection. Thus

$$D \subseteqq (B \cap C) \subset A)$$

[4]) Or at least it is straightforward when it is answered by *ourselves* who have created those systems. Whereas a visitor from another planet might not know what is the function of human devices or institutions until such time as he had learnt to understand the human "life style". Thus "function" is perhaps not fully specifiable in the sense of being able to program a computer to recognize it.

[5]) Some valuable insights concerning the extent to which the operational principles of machines are not entailed by the laws of physics and chemistry are due to M. Polanyi—for example in his *Personal Knowledge* [9] or *The Tacit Dimension* [10].

[6]) Some useful discussions concerning "organization" and "function" are those of Waddington and Caspari [11; 12; 13].

Integrality

As has been said already an organized system is necessarily an assembly of parts and sub-parts etc. which are interconnected.[7]) It is the connectedness of the assembly which gives to it an essential unity.[8]) On the other hand not every connection between the various parts is necessarily equally useful. As a first approximation, distinction needs to be made between: (1) those types of connections which facilitate the function, or functions, of the whole assembly; (2) those which operate against these functions; and finally (3) those which are irrelevant.

A further useful approximation is to suppose that each of the component parts has a fixed and severely restricted capacity. Each can contribute only a portion to the overall performance. Therefore if a system such as a living organism or a business or a machine is to be capable of a variety of functions it must contain *many different sorts of parts*. Thus in regard to the quality of being "highly organized" it is probably more essential for a system to have a big variety of components than for any of them to be duplicated (although the latter may be very important in regard to reliability of performance).

Let degree (or amount) of organization be called *integrality* for short and be denoted by the symbol φ. If the above ideas are on the right lines it would appear that

$$\varphi = f(c, n, x_1, x_2, \ldots, x_n) \tag{1}$$

where c is the number of connections which are of the first type mentioned above, n is the number of different sorts of parts in the above sense and x_1, x_2 etc are other variables determining the value of φ which go beyond the scope of the present tentative discussion. It would be reasonable to suppose that φ is a monotonously increasing function of c and n, but not necessarily of the other variables.[9]) Thus a *possible* form of the function is

$$\varphi = cnf'(x_1, x_2, \ldots, x_n) \tag{2}$$

where φ is here assumed proportional to the product of c and n.

The value of φ must, of course, be reckoned relative to some arbitrarily chosen base line. For instance in reckoning the integrality of, say, a clock one might reasonably take the base line at the level of the wheels, axles, springs etc. and these would therefore be regarded as

[7]) From this point of view the theory of organization may be treated as a branch of graph theory.

[8]) The unity of organized systems has naturally been given great emphasis in holistic theories. A closely reasoned account of the nature of "wholes" has been given by Marjorie Grene [15, p. 358]. It is the whole, as she says, which defines the relevant parts; to be sure the parts are the *conditions* for the whole which could not exist without them; "but it is the whole that *explains* the parts, not the parts the whole. The whole is the system (the organism) that makes the parts the parts that they are, even though the parts are the conditions (in the traditional language of material causes) for the existence of the whole".

[9]) For instance φ might be a decreasing function of the number, x_1, of connections which operate *against* the performance of the assembly. Other important variables, x_2, x_3 etc. might be weighting factors which describe the relative significance of various connections or of various components.

having no integrality of their own.[10]) Even so it should be noticed how rapidly the product cn increases with increasing complexity of the system. If the latter contains a total of N components (where N, unlike n, includes all replicas), the minimum number of connections which will link them all together is N-1, whilst the maximum number of connections, that which would result from a two-way linking of every component to every other, is $N(N\text{-}1)$. Thus for large values of N the number c of connections lies approximately between N and N^2, according to the degree of completeness of the connective system. If this is now multiplied by n, which is a measure of the distinctivity of the N components, one would obtain a product which in the case of, say, an aircraft would be an extremely large number and in the case of a living creature, on any reasonable choice of the base line, would be quite astronomically large. These considerations are clearly in line with von Neumann's theory [6].

Although the foregoing touches only the fringe of the problem the main point I want to stress is that the means of making the concept of an organized system more precise lies in the direction of linking it to the more readily quantifiable notions of degree of connectedness and distinguishability of components. When a living creature becomes more highly organized during ontogenesis this must be due, at least in part, to: (a) increasing number of channels of internal communication, including feed-back channels; (b) increasing number of distinctive organs and tissues. That is to say to the factors c and n respectively, as already discussed.

Integrality as Non-Conserved

If the connectedness of an organized system has the significance which has been suggested an important conclusion may be drawn. This is based on the fact that there is no conservation law which prevents changes in the number of connections. Such changes might occur, for example, through the movement of component parts into, or out of, direct physical contact, or they might be due to the switching on or off of circuits, or to the blocking or unblocking of channels. Changes of this sort are in no way forbidden by any of the known conservation laws. Thus if the integrality of a system depends on the factor c there is no requirement from physics that this factor shall remain constant during the course of time even if the system is isolated. It might either increase or decrease.

An instance of increasing integrality is provided by the example of a bird's egg. An egg, because of its hard and almost impermeable shell, is almost a closed system.[11]) An egg, together

[10]) As noted already I have not attempted an answer to the question whether atoms and molecules should be regarded as being organized systems, or whether "organization" begins at a much higher level, for instance at the level of cells. Fortunately this uncertainty does not invalidate the main point of the article concerning φ as being non-conserved.

[11]) See, for example, Chr. Raven [16]. The same example of the bird's egg serves to show that 'organization' bears no immediate relationship to "information"—i.e. in the sense of information theory. The amount of information required for the embryo's development is assumed by many biologists to be contained entirely within the egg and is conserved. Thus if the amount of Information remains constant during a period when integrality is increasing, these two attributes obviously cannot be the same.

with a suitable incubator, offers an example of an isolated system. Nevertheless it can hardly be doubted that, during the gradual development of its embryo, the degree of organization of such a system greatly increases.[12])

If integrality is thus a non-conserved quantity, so also of course is entropy. This raises the question whether a change in the one quantity is *necessarily* related to a change in the other. That this is not the case may be seen from either of the following considerations: (1) the previously discussed absence of any direct relationship between organization and orderliness; (2) more rigorously from the fact that the making or breaking of connections could occur reversibly in a suitably idealized and friction-free system, thus resulting in changes of integrality at constant entropy.[13])

This is not to say, of course, that there is not often some degree of linkage between processes leading to change of integrality and those leading to change of entropy. The growth and differentiation of an organism, together with all consequential changes of its foodstuffs and environment, must surely be assumed to be entropy producing, even though the actual evidence is scanty.[14]) What *is* being said is that the linkage, when it occurs, is not a necessary one and this is the significant point.

At the beginning of the article it was suggested that any non-conserved quantity arising from the field of the life sciences needs to be independent of thermodynamics and should also have considerable generality. The matter must now be considered from the viewpoint of this second criterion.

The non-conserved aspect of integrality displays no great significance, I think, in connection with any *single* organism. Presumably the integrality of the organism rises to a maximum during youth and maturity, and subsequently falls away rapidly as death is approached. Thus there can be no suggestion here of any monotonously increasing behaviour.

Consider however the *total* integrality, denoted Φ, of the Earth's ecosystem. Let φ_i be the integrality of any single organized entity, of the i'th type, and let p_i be the population of all such entities within the ecosystem. If they may be assumed for simplicity to be independent of each other[15]) their total contribution to the integrality of the ecosystem will be the product $p_i\varphi_i$. Similarly Φ will be obtained by summing $p_i\varphi_i$ over all types i. Thus:

$$\Phi = \Sigma p_i\varphi_i. \tag{3}$$

[12]) From the same point of view, damage to an organism which results in death may be regarded as giving rise to a sudden decrease of the organism's integrality, due to the breakdown of essential communication channels.

[13]) The conclusion that degree of organization and entropy are not related was also drawn by M. J. Klein [17].

[14]) Some calculations based on experimental data on cell growth are described by Linschitz in *Information Theory in Biology* [18].

[15]) The entities considered here are the various organisms. In so far as, (a) they are not independent but form with each other 'societies', or (b) they construct various organized structures external to themselves, these two factors would tend to cause the equality sign in (3) to be replaced by a "greater than" sign.

It may be noted too that it is assumed that the integrality of all the various organisms is reckoned relative to the same base line—e.g. cell components taken as zero.

Of course it must be admitted that the possibility of actually evaluating any of the φ_i is exceedingly remote; indeed the concept of integrality which I have been advancing is schematic rather than practical in its intention. Nevertheless if the theory underlying equation (3) is reasonable, is the conclusion not almost inescapable that Φ is an increasing function?

I mean of course since life first began on this planet.[16]) For in the first place the process of evolution has been one in which life-forms of increasing degree of complexity have made their appearance successively. From a beginning which was non-organic there have emerged the single-celled organisms, the more highly organized multi-cellular creatures and those having highly differentiated organs and tissues, and finally the human brain with its thousands of millions of interconnected neurones. Thus it would appear that many of the individual values φ_i, attributable to each member of the various species, have greatly increased.

Φ must be regarded as having risen greatly in value for a second reason: because many of the population terms p_i have also increased and this applies especially to the most complex organisms, those having the highest φ_i values, due to natural selection. Of course it is true that a number of species have disappeared. Yet the significant point is that the majority of species whose members had a high intrinsic degree of organization also had increasing values of their populations p_i. It is on grounds such as these that it seems reasonable to suppose that the total integrality Φ of the earth's ecosystem has greatly increased and is probably still increasing.[17])

No doubt many evolutionists and biologists have had some such view in mind. For example J. Needham has written: "The law of evolution is a kind of converse of the second law of thermodynamics, equally irreversible but contrary in tendency" [20]. Unfortunately there has also been a school of thought which supposed that organisms somehow succeed in contravening the Second Law of Thermodynamics. What I have sought to show in this paper is that there need be no contravention. Organisms may well be fully subject to the Second Law, manifesting entropy increase like any other material systems[18]), and yet display a temporal trend which *is entirely different and independent*. The significant points are that the factor c in equation (1) is non-conserved and that consequential changes of integrality bear no necessary relationship to changes of entropy. Changes of φ and Φ are therefore outside the scope of the Second Law and *are in no way contrary to it*. It would appear, in short, that the most significant processes which are studied in biology, the evolutionary processes, have no direct connection with thermodynamics.

[16]) Maynard Smith [19, p. 266] also reached the conclusion that considerations of increasing complexity should be applied, not to single individuals or even to single species, but rather to the entire ecosystem. He writes: "If there is a 'law of increasing complexity', it refers not to a single species, ... but to the ecosystem as a whole. The complexity of the most complex species may increase, but not all species become more complex".

[17]) Earlier in the paper I included certain non-living entities within the concept of organized systems. For example, all of mans' material constructions; also his abstract constructions such as languages, legal codes, scientific theories etc. Allowance for these within the calculation of Φ still further confirms the notion of Φ as an increasing function within the Earth's ecosystem.

[18]) One should allow, of course, for the fact that organisms are open systems. The Second Law must therefore be applied to the total system, organism plus environment.

Of course this does not imply vitalism. Good physical reasons can be given for changes of integrality without invoking anything which is unknown. But the direction of such changes, whether towards an increase or a decrease, must probably be accounted for by means of boundary conditions rather than laws. The forces which can establish connections in material systems are well-understood (e.g. electrostatic, covalent or Van der Waals forces) but it cannot be predicted whether a given connection will be formed or broken without additional knowledge concerning the boundary conditions under which these forces act. Processes of dis-aggregation take place, as well as processes of aggregation, and the circumstances which determine the actual outcome are left open by the physical laws. Thus it would appear in the present context that the combined effect of the known physical forces and of the actually prevailing conditions in the biosphere must be such as to give rise to an overall connection-forming tendency in organized systems.

Of course it is very far from my intention to suggest that there can be a law of increasing integrality (even if my definition of this quantity were much less rudimentary than it is) which could have anything approaching the degree of universality of the law of increasing entropy. Obviously enough a catastrophe of a sufficient magnitude could result in the termination of life. The concept of integrality may perhaps be applicable on a universe-wide scale, but it will remain quite unknown whether Φ, if it were summed over all planets where there are organisms, is increasing or decreasing. Life may be disappearing on some planets, due to the dying out of their suns, faster than it is appearing on others. To suppose otherwise would be an act of faith! Therefore from a purely scientific viewpoint I would regard the notion of a *specifically increasing* integrality as having no more than a very limited scope – i.e. as being applicable to the Earth's species over a certain duration of time. But in this it is no worse a case than other biological "laws"!

In conclusion I would like to offer my best thanks to Professor G. Chapman for his very helpful comments on an early draft of the paper.

References

[1] P. Weiss in Biological Organization (C. H. Waddington, Ed.). Pergamon, Oxford 1959.
[2] P. B. Medawar: The Herbert Spencer Lecture, Oxford. Encounter *21* (September 1963), 3, 35—43.
[3] E. Schrödinger: What is Life? Cambridge University Press, Cambridge 1951.
[4] B. C. Goodwin: Temporal Organization in Cells. Academic Press, London 1963.
[5] K. G. Denbigh: The Principles of Chemical Equilibrium. Third Edition. Cambridge University Press, Cambridge 1971.
[6] J. von Neumann: Theory of Self-Rerpoducing Automata. University of Illinois Press, Urbana 1966.
[7] I. Prigogine: Time, Structure and Entropy. In: Time in Science and Philosophy (J. Zeman, Ed.). Academia and Elsevier, Prague and Amsterdam 1971, 89—99.
[8] I. Prigogine and G. Novalis: Biological Order. Quarterly Review of Biophysics *4* (1971), 2—3, 107—148.
[9] M. Polanyi: Personal Knowledge. Routledge and Kegan Paul, London 1958.
[10] M. Polanyi: The Tacit Dimension. Routledge and Kegan Paul, London 1967.

[11] C. H. Waddington: The Strategy of the Genes. Allen and Unwin, London 1957.
[12] C. H. Waddington: Biological Organization. Pergamon, Oxford 1959.
[13] E. Caspari in Frontiers of Science and Philosophy (R. G. Colodny, Ed.). Pittsburgh University Press, Pittsburgh 1962.
[14] J. H. Woodger: Biological Principles. Routledge and Kegan Paul, London 1929.
[15] M. Grene: The Knower and the Known. Faber and Faber, London 1966.
[16] Chr. Raven: Oogenesis: The Storage of Developmental Information. Pergamon, Oxford 1961.
[17] M. J. Klein; British Journal for the Philosophy of Science *4* (1953), 14, 158—160.
[18] Linschitz in Information Theory in Biology (H. Quastler, Ed.). University of Illinois, Urbana 1953.
[19] J. M. Smith; Studium Generale *23* (1970), 266—272.
[20] J. Needham: Time: The Refreshing River. Allen and Unwin, London 1943.

PROBLEM OF DIRECTION OF TIME AND THE LAWS OF SYSTEMS' DEVELOPMENT

A. I. Uyemov

Science in the 20th century has known many discussions on many a seemingly evident theorem which goes to prove that science does not satisfy itself with the external factors of phenomena but makes every effort to find their substance. One of such discussions is that on the problem of the direction of time. Why is time irreversible? What are the objective criteria for determining the future and the past? These questions lead us from the generally known facts of everyday life to the analysis of the fundamental laws of space.

Various replies may be applied to the above posed questions. Their analysis is not the aim of this paper. Nevertheless we should like to mention briefly one of them. Starting with Leibniz and Kant, attempts have been made to explain the asymmetry of time by the asymmetry which exists between the cause and the effect. Experiments of this type are usually termed the "causal theory of time". The theories summed up under this term have all encountered basic difficulties, the most serious of them being that of the danger of the vicious circle. This results from the fact that the relation of precedence in time is determined as a relation. Here we should mention the criticism of the causal theory of time by H. Reichenbach as presented by A. Grünbaum [1, p. 237]. The mentioned difficulty can be eliminated if we abandon the concept that cause precedes effect in time. J. Whitrow discussing Kant's causal theory of time notes that Kant's difficulty was in that the most effective natural causes appear simultaneously with their effects [2, p. 349]. Insofar as our point of view is concerned it is not so much that these causes appear simultaneously with their effects but that they occur simultaneously. There is ample written evidence by authors who hold that causes and consequences occur simultaneously [3; 4; 5; 6; 7; 8].

From this point of view the criterion for *determining* the direction of time can be defined by using the notion of accumulation of effect. "If the accumulation of effect in the direction from t_α to t_a is linked with the weakening of the cause then t_α precedes t_a" [9].

Another trend of thought, insofar as the problem of the direction of time is concerned, is linked with the application of the second theorem of thermodynamics. It appeared at the beginning that in a closed system entropy can only grow and, therefore, the entropy increment would show the direction of time. The discoveries made by L. Boltzmann, to the effect that this law has a statistic character made the situation only more difficult. The analysis made by H. Reichenbach showed that the curve of entropy increment characterizing one isolated system does not give the basis for any conclusion on the direction of time [10, p. 160]. Reichenbach found that the solution lays in departing from the study of one isolated system to the study of a multitude of systems, a set of systems in space. In such a set of systems the direction of time may be determined according to the changes in entropy characterized by most systems in the given set. It does, however, still remain unclear why all "branching systems" which are elements in the space set cannot be conceived as elements of one system.

Reichenbach uses the terms "one system" and "many systems" at the intuitive level without any precise definition. His conclusion is, therefore, not conclusive. This shortcoming is also true of the way in which Reichenbach and others apply the term entropy. Yet, there does exist an adequately clear explanation of this notion presented by L. Boltzmann, i.e. the logarithm of the probability of the state. There does also exist the algorithm of calculating the probability of the state in that case when the system represents the final set of molecules which are situated at mutually equal points in space. This algorithm does not make it possible to determine that the entropy of a broken glass will be larger than that of an unbroken glass. When Reichenbach and other authors say that in such cases there is an increment of entropy they conceive this term in a very general sense of the word as one defining a certain measure of the lack of organization manifested in the system. In such a case the statement on the growth of entropy assumes the character of a philosophical concept on the growth of disorganization. It is considered adequate that this concept is equivalent to the second formula of thermodynamics. Generally speaking this is not so because the concept of disorganization then becomes too indefinite. The question of the equivalence of the mentioned formulas can only be presented clearly if the notion of organization is clearly defined. The analysis of texts in which this term is used shows that it is being used in different contexts. Reichenbach, for instance, often conceives it in the subjective sense and it is, in fact, identified with the concept of usefulness for man. The glass that falls on the floor will break and one cannot drink from a broken glass. The subjective criterion gives no prospect. In some cases it will be just that broken glass which we shall find useful. And insofar as walnuts are concerned they are useful for man and other beings only in a broken state.

If we conceive organization as the existence of differences between the significance of the variables that characterize the individual parts of systems (e.g. the existence of the temperature gradient), independent of the needs of man, then the shattering of the glass can also be explained as diminishing entropy; the breaking of the glass may also increase the distance separating the individual parts of the system from its centroid. If we paste the bits of glass together we reduce this distance and thus, according to the given definition of organization, we increase entropy.

In the discussions which scientists held immediately after Clausius had set the second theorem of thermodynamics, many examples were presented of the spontaneous growth of distinctions, including the increment in temperature differences. If, for instance, we have oxygen and hydrogen of the same temperature in two different parts of one vessel, separated by a porous wall, then the difference in temperature arises at the cost of a larger diffusion of the more rapid molecules of hydrogen from one section of the vessel into the other. This conceptual experiment was presented by T. Preston as an argument against Clausius [11, p. 17]. The latter in his reply pointed to the fact that the difference in temperature occurs at the cost of the diffusion of gases inside the vessel, i.e. by the elimination of another previously existing difference.

Other phenomena of a similar type could be mentioned. The explanation of such phenomena from the point of view of the second theorem of thermodynamics has resulted in the generalization of the notion of entropy.

However, all cases where the growth of organization in the system is accompanied by the growth of disorganization in the broader system can also be studied from another aspect, namely, that the *origin of disorganization in one system is linked with the growth of organization in another system*. We shall term this concept the *principle of negative entropy*.

A vast number of examples may be presented at this point to document the correctness of the principle of negative entropy. Thus, for instance, the disorganization of street traffic leads to the establishment of order, automobiles line up at the crossroads waiting for the green light. It is disorganization in the selection of investigated objects that ensures the organization required by the principles of inductive conclusions.

The principle of negative entropy in its generally philosophical concept has the same right of existence as the principle of the growth of entropy.

Both principles can be linked up in the principle of the conservation of entropy as the measure of disorganization which, together with the theorem of the conservation of energy and the theorem of the conservation of mass, can be conceived as as a partial case of the general principle of conservation as it was formulated by M. V. Lomonosov: "All changes occurring in nature take place in such a manner that insomuch as is added to one phenomenon, is taken away from another." [12, p. 160].

The correctness of the law of conservation of entropy can be represented in a definition if the adequate definition of organization and entropy is applied as the measure for the degree of the absence of this organization.

The correctness of the principle of the conservation of entropy means that the growth of entropy cannot be applied for determining the orientation of the direction of time. The transition to the space set (Reichenbach) will not help us. In case of a sufficiently large number of systems in such a set we shall have approximately an equal number of systems with a diminished entropy and systems with an increased entropy.

Summing up what we have said so far we may say that attempts to link the direction of time with the basic principles of physics have not been successful. This is certainly not a chance phenomenon. The main cause is evidently the fact that the very approach to the matter is not adequate, i.e., that the endeavour of the research workers to use general concepts, be they only partially scientific, to attain a solution of a problem related to the entire world, has not been adequate. Time is a form of the existence of matter [13, p. 350]. This means that "to reduce time to a mere physical parameter would be making the same error as was that made by those who in their time reduced motion to mere mechanic motion", as has correctly been stated by F. Askin [14, p. 168].

What then is the alternative to a partial scientific approach? Numerous scientific explorers have united many partial scientific approaches. J. Whitrow, for instance, does not restrict himself to physics but also studies time from the point of view of biology, psychology and sociology. In this way, however, the unity of time disappears. This unity could have been secured by the systems approach in general, independent of their physical character. The scientific branch which is called the systems approach or "systemology" is now being established. Its predecessor used to be "tectology" by A. Bogdanov and "the general theory of systems" by L. Bertalanffy. Current variants of system research can be found in many materials [15; 16; 17; 18; 19; 20; 21].

Let us note, first of all, that the problems which we have mentioned above have a system character. It is the problem of the laws of the development of systems and the problem of the unification of various systems into one system.

The growth of entropy may be used for determining the orientation of the direction of time only insofar as this increment is not conceived as a partial physical theorem but as the theorem of the development of random systems, i.e., as a system theorem. We have attempted above to prove that such procedure is not adequate.

There can, however, exist other general systems principles which will be used for the solution of the set tasks. Before we formulate them, we have to analyse certain notions pertaining to the systems approach. Let us start with the notion "system" proper. The usual definition of the system as a summation of elements that are mutually effective and mutually linked is somewhat narrow despite the fact that it is currently generally recognized. It does not take into consideration such systems as, for instance, is the natural order of numbers or the Mendeleyev system of elements. It is true that one can say that such definitions are completely adequate insofar as the determination of the time vector is concerned, but as for the system of the natural order of numbers one cannot speak of time at all. Numerous systems are really determined independent of time and equally independent of movement, so that in relation to them we cannot even introduce the notion of state. Nevertheless, the general aspect always has its adavantages even for studying such partial cases for which it is not indispensably significant.

Let us now define the notion of the system as a multitude of elements (m) in which the relation (R) is realized with the previously fixed property (P) [22; 23]. The formula is as follows:

$$(m)\ S =_{\mathrm{df}}[R(m)]\ P.$$

The so-called systems parameters are conceived as the specific systems properties [24; 25] and their significance of its substrate. Such parameters include, for instance, the possibility or impossiblity of increasing the number of elements in a system. Such qualities as mass, volume etc., are not systems parameters. Parameters may be dual (the above mentioned case), ambiguous and linear. To mention an example of the linear parameter let us recall the simplicity and complexity of systems.

Also organization and disorganization can be considered as a systems parameter. As has been mentioned above this parameter has not been defined clearly enough. The introduction of the measure of disorganization, i.e. entropy, has, it appears, clarified this parameter. It appears that this parameter is clearer when the measure is represented numerically, e.g., in the theory of information. This is true, for instance, of the Shannon formula. Contrary to thermodynamic entropy, entropy defined according to the Shannon formula, may be conceived as a sufficiently accurate measure of the systems parameter. Will this be the parameter of organization and disorganization? The very simple examples we have mentioned give us reason to doubt this. Entropy of replies linked with the throwing of an even-shaped dice will be larger than entropy linked with an unevenly-shaped dice, e.g., one whose one half will be heavier than the other. The intuitive evaluation of the measure of organization will be larger in the case of the regular dice than in the case of the irregular dice. Entropy, according to Shannon, is thus not a measure of disorganization but of another *systems parameter*, namely the *substrate homogeneity* of the system. This can mutatis mutandis be said of other measures observed in information theory.

The question naturally arises whether, if it is not possible to apply for determination of the time arrow the parameter of organization, it is not possible to apply the parameter of homogeneity which has an accurate numerical representation. The reply to this question depends on whether or not there exists a general law related to the growth of entropy as a measure of the homogeneity of systems. There are apparently very few reasons to assume the existence of such a principle. Many examples could be put forward to support and to oppose it. Thus, for instance, mountains get smaller by weathering while valleys are filled with sediments. On the other hand, the landscape becomes rugged by canyons. In his time H. Spencer formulated as the general law of the world and its development the trend of transition from homogeneity to heterogeneity [26].

In his criticism of Spencer Mikhaylovsky attempted to prove with many examples that what was homogeneous could not by itself become heterogeneous and, that there existed a universal law of the transition from a lower degree of heterogeneity to a higher degree of heterogeneity [27].

In both cases, i.e. according to Spencer and to Mikhaylovsky, the universal law would be, contrary to our assumption, the law of the diminishing of entropy as the measure of the homogeneity of systems.

It is evident that the problem of what trend (whether towards homogeneity of systems or to heterogeneity of systems) is predominant, cannot be solved at the level of empirical cognition.

Apart from the above mentioned parameters there does exist, as we have also mentioned above, a number of other systems parameters. Basically, each of them could be applied for determining the orientation of the time vector and such application would be equally justified. It is interesting that here again we can apply the measure of entropy. By applying the measure of entropy we are able to evaluate the simple or complex character of systems as the meaning of one of the linear system parameters [28; 29]. Nevetherless, empirical data do not allow us to determine unambiguously whether or not the respective system tends to develop into a complex or into a simple one.

Other parameters, especially binary systems, cannot be linked with the notion of entropy. Here the object is to analyse the laws of the system development which are related to these parameters on that level at which the analysis had been conducted regarding the entropy as the measure of disorganization. As this, however, is the empirical aspect of the law the analysis exceeds the scope of this paper. There is no reason to expect that empirical data will prove with more or less precision this or that law.

Let us now look into the theoretical aspect of the problem. L. Boltzmann did not found his formulation of the relation between the growth of entropy and the direction of time merely on empricial data but more on the belief that, disorganization is more probable in systems than is organization and, that systems in themselves attempt to transcend from the less probable state to the more probable one. Both these assumptions can be doubted. However, irrespective of whether the assumptions are correct, their probability does not lead us to the required conclusions.

Let us now present the possibilities of the solution of the investigated problem at the theoretical level. The analysis of systems parameters shows that their logical structure can be represented by means of the classification of things, properties and relations. Such representations can be made with the help of a special system created for the purpose [30].

The classification of "things", "properties", "relations", makes possible the approach by which difficulties are pinpointed and overcome. It can be said that the common feature of all attempts at determinination of the time arrow has been the determination of the asymmetric relations between kinds, especially between processes. It has always been proved that the asymmetry has a statistical character which proved inadequate for the solution of the problem.

Another way towards the solution of the *problem assumes the transition to relations of another order, i.e. between things on the one hand and relations which exist in these things on the other.*

It is here that the asymmetry we are seeking can be found, such as no longer has a static but a dynamic character. As the object of investigation let us choose a relationship of another order, one which can roughly be termed the effect or the impact. And it was this type of relation that was the subject of discussions on the philosophical aspects of the theory of relativity when the mutual effect of time-space continuity and mass [31] was emphasized.

To represent this problem let us use the notion of internal and external relations which has already been introduced [32; 33].

The internal relation between objects $x_1, \ldots, x_n$ is viewed as a relationship which is indispensable for the existence of these objects. The external relation is then determined not only by the related objects themselves but also by certain conditions depending on external objects. For instance, the relation of "older than" which exists between Peter and Ivan is an internal relationship between two people while the "higher than" in the sense of "he holds a higher position at his job" is an external relation.

How can things $x_1, \ldots, x_n$ affect relation R which exists between them? Two cases can occur. In the first case relation R is an inner relation for its correlates. In this case things $x_1, \ldots, x_n$ may change R but, their mutual relation only by transforming their very selves. In the second case R is an external relation for $x_1, \ldots, x_n$. In this case the relation can only be transformed by the influence of things $y_1, \ldots, y_k$ which are external in relation to the set $x_1, \ldots, x_n$. In such a manner it is possible to formulate the following statement: If things in a system do not change the relations between them can only change as a result of some external effect.

Let us attempt to formulate a parallel statement: "If the relations between things in a system do not change them the things themselves cannot change otherwise than by the effect of some external factor."

And yet, this statement will not be correct. The principle of pairing which is valid for pairs of categories such as: "property-relation", "thing–property", is not valid for pairs such as "thing–relation". And *it is here that we find the asymmetry* which will help us define the orientation of the time vector.

Let us now return to the formulation which I found as incorrect. Things existing in given relations can change as the result of the functioning of the given relations, independent ofany external effect. If some relationship is established between people, e.g., the relation of management, then this relation will transform people regardless of whether the relation itself remains unchanged. This very property of relations was taken into account at the concluding of marriages according to the principle "if they live together–they love one another".

Not only people but also inanimate things get used to one another, more precisely, they become used to mutual relations. A rubber ball which gets deformed by being bashed against a wall, rebounds and recovers its form. However, if we press a rubber ball against a wall and keep it in that position for several years, the rubber ball will become adapted to its new position. The relation has not been transformed but the rubber ball has.

The currently applied experiments aimed at the determination of the time arrow are linked with the generalization of the results of thermodynamics. We shall, however, base our observations on another discipline. One that has recently become an independent discipline. It is rheology, i.e. the science related to deformations and the flow of matter. The concept of "they live together–they love one another" is expressed in the following manner in rheology: If we take a brief look at the results of experiments studying the effect of the previous discharge pressure on the consequent deformation of thrust, pressure and chute we find that the creep limit in repeated thrust under atmospheric pressure is considerably lower than in the previous deformation but, that the inclination of the curve showing consolidation under pressure is higher than had been the inclination of the respective curve in the previous deformation under atmospheric pressure [34, p. 371].

We have explained the asymmetry that occurs in the relation of another order between things and relations. As another step on the way to finding a solution of the studied problem let us introduce the notion of the "relational collapse". As the model we shall use here the concept of the gravitational collapse. In the case of the latter an arbitrary material entity, including photons, cannot leave a certain sphere (zone) once it has entered this sphere. The relational collapse means that once things enter a certain relation they reach such a state (resulting from the effect of this relation) that they cannot get rid of this relation otherwise than by abolishing their existence as given things.

The gravitational collapse can be conceived as a partial case of a relational collapse. The rubber ball which has lost its flexibility as a result of lying on the table for many years will be an example of a thing in the state of a relational collapse insofar as the relations of the concerned. The ageing of the organism is evidently linked with the relational collapse into which the elements of this system enter. Numerous examples presented as non-thermodynamic causes for the anisotropy of time can be conceived as relational collapses [2, pp. 16–20; 1, pp. 327–345].

Mathematically, the relational collapse is represented as a function to whose values (ranging from the certain value of the argument) certain limitations are assigned.

Let us now transform the above concepts into the language of systems parameters, with whose help it is possible to express the general laws of systems which determine the orientation of the time arrow.

Let us set the following limitations:

The systems in which the system constituting relation is inner for their substrate will be termed as *internal*, systems of the opposite type will be termed *external*. As an example of the internal type let us give a set of people united by the age relation which can be "older than". Serving as an example of the external system will be a set of people occupying a certain position in the employment hierarchy. The internal aspects form two aspects of a certain systems parameter.

The formula of the internal systems in the language of the ternary description [30] will be as follows:

$$(a)\ S_{\mathrm{int}} =_{\mathrm{df}} [a\,(\neg\, a)]\, t \dashrightarrow [\neg\, a \dashrightarrow \neg\, a(\neg\, a)].$$

S_{int} in the metalanguage means the abbreviation for "to be the internal system". The first bracket expresses existence in a determined system, in the same to which the property S_{int} is being assigned, a property which has a certain relation a which has a previously fixed system property t. In this case we obtain the second bracket telling that if there exists the same substrate that has been mentioned before then there also exists the same systems formative relation which is materialized in this substrate.

A theorem can be proved according to which *substrates of internal systems, in which the introduction of the notion of the state has meaning, exist in the conditions of the relational collapse.*

Let us base our premise on the opposite assumption. If the substrate has not existed in the conditions of a relational collapse then this would mean that the elements of the substrate can mutually affect one another in a manner which negates that which has been fixed by the systems formative relation R. The transition from relation R to relation $\neg R$ which negates it can be realized only by the functioning of external or internal factors. However, the external factors cannot by themselves change relation R insofar as it remains internal for its correlates. External factors can affect relation R only through the transition of the correlates of this relation. If the change of correlates of the internal relation can make that the relation ceases being internal then a way will have been found of changing the relation without changing the correlates themselves. This, however, is adverse to the very notion of the internal relation itself.

The proved theorem shows that *if the external system changes into an internal system then this system remains an internal system.*

We have thus set as the general law of systems the law of the transition from the external system to the internal system. On the basis of general speculations we cannot say what system will change from an external one into an internal one and when. This is a matter of statistics. Adverse transitions are not a matter of statistics as they are excluded by the preceding conceptions.

From this ensues the possibility of determining the orientation of the time vector, namely, *time flows in that direction which corresponds to the transition of external systems to internal systems.*

We have studied one parameter and one general law of the systems approach linked with this very parameter which can be used for determining the flow of time. It is evident that other parameters can be applied analogically.

The following question does, however, always remain to be solved: if the direction of time is linked with the transition of one type of system to another, how is it then that space has not yet reached such a state in which there are only systems of one type. A similar problem, namely that of heat death arises in the approach to the orientation of the time vector from the entropy point of view. There, the situation is mitigated by a multitude of adverse transitions connected with diminished entropy. Other physical and philosophical considerations could be presented to oppose heat death. In our case the problem is solved in a different manner. The transition from external to internal systems is related to the development of systems. Together with the systems development we have the origin, genesis of new systems. The notion of genesis in many aspects, mainly in the systems approach, differs considerably from the notion of development [35].

The developing systems can be both external and internal. The former transcend into the latter. Simultaneously with this originate new external systems, etc.

References

[1] А. Грюнбаум: Философские проблемы пространства и времени (A. Grünbaum: Philosophical Problems of Space and Time). M. 1969.

[2] Дж. Уитроу: Естественная философия времени. (G. Whitrow: The Natural Philosophy of Time). M. 1964.

[3] A. Ujemov: Předchází příčina účinku? Filosofický časopis (1959), 6, 836—848.

[4] A. Ujemov: Spor o slova a reálný problém. Filosofický časopis (1961), 6, 903—905.

[5] А. И. Уемов: О временном соотношении между причиной и действием (A. I. Uyemov: On the Time Relation between Cause and Effect). Иваново 1960.

[6] А. Поликаров: Относительность и кванты (A. Polikarov: Relativity and Quanta). M. 1966.

[7] Г. А. Свечников: Причинность и связь состояний в физике (G. A. Svechnikov: Causality and Connection of States in Physics). M. 1971.

[8] М. А. Парнюк: Принцип детерминизма в системе материалистической диалектики (M. A. Parnyuk: Principle of Determinism in the System of Materialistic Dialectics). Киев 1972.

[9] А. Уемов: Причинная связь и необратимость времени (A. I. Uyemov: The Causal Connection and Time Irreversibility). In: Проблема причинности в современной физике. Гл. Ш. M. 1960.

[10] Г. Рейхенбах: Направление времени (H. Reichenbach: The Direction of Time). M. 1962.

[11] Ф. Розенбергер: История физики. Часть третья, выпуск II (F. Rozenberger: History of Physics. Vol. III, 2nd issue). М.—Л. 1936.
[12] М. В. Ломоносов: Избранные философские произведения (M. V. Lomonosov: Selected Philosophical Works). М. 1950.
[13] Ф. Энгельс: Диалектика природы (F. Engels: Dialectics of Nature). К. Маркс, Ф. Энгельс: Соч. 20. М. 1961, стр. 550.
[14] Я. Ф. Аскин: Проблема времени в физике и диалектическая концепция развития (Ya. F. Askin: The Problem of Time in Physics and the Dialectical Conception of Evolution). In: Пространство и время в современной физике. Киев 1968.
[15] Проблемы формального анализа систем (Problems of Formal Analysis of Systems). М. 1968.
[16] Исследования по общей теории систем (Research in General System Theory). М. 1969.
[17] Проблемы методологии системного исследования (Methodological Problems of System Research). М. 1969.
[18] Системные исследования. Ежегодник 1969 (System Research. Annual Vol. 1969). М. 1969.
[19] Системные исследования. Ежегодник 1970 (System Research. Annual Vol. 1970). М. 1970.
[20] Системные исследования. Ежегодник 1971 (System research. Annual Vol. 1971). М. 1971.
[21] Промышленная кибернетика" (Industrial Cybernetics). Киев 1972.
[22] А. И. Уемов: К вопросу об определении понятия „система" (A. I. Uyemov: On the Question of the Determination of Concept „System"). In: Некоторые теоретические вопросы коммунистического строительства в СССР. Одесса 1967.
[23] А. И. Уемов: Системы и системные исследования (A. I. Uyemov: Systems and System Research). In [17].
[24] А. И. Уемов: Системы и системные параметры (A. I. Uyemov: Systems and System Parameters). In [15].
[25] Г. Я. Портнов, А. И. Уемов: Системные параметры и физические величины (G. Ya. Portnov, A. I. Uyemov: System Parameters and Physical Quantities). In: Філософські проблеми сучасного природознавства. Выпуск 22. Київ 1971.
[26] Г. Спенсер: Сочинения. т. I. Основные начала (H. Spencer: Works, Vol. I. First Principles). Санкт-Петербург 1899.
[27] Н. К. Михайловский: Что такое прогресс? (N. K. Mikhaylovsky: What is Progress?) Петроград 1922.
[28] А. И. Уемов: Упрощающие свойства отношений и меры простоты систем (A. I. Uyemov: Simplifying Properties of Relations and Measures of System Simplicity). In: Філософські проблеми сучасного природознавства, Випуск 27. Київ 1972.
[29] А. И. Уемов, Б. В. Плесский, Л. Н. Сумарокова: Информационные процессы в научном исследовании и проблема их упрощения (A. I. Uyemov, B. V. Plessky, L. N. Sumarokova: Information Processes in Scientific Research and the Problem of their Simplification). In: Проблемы информатики. Выпуск 3. Новосибирск 1972.
[30] А. И. Уемов: Выражение определения системы и системный параметров на языке тернарного описания (A. I. Uyemov: Expression of System Determination and of System Parameters in the Language of Ternary Description). In [21].
[31] А. И. Уемов: Может ли пространственно-временной континуум взаимодействовать с материей? (A. I. Uyemov: Can the Space-Time Continuum Interact with Matter?) Вопросы философии (1954), 3.
[32] А. И. Уемов: Вещи, свойства и отношения (A. I. Uyemov: Things, Attributes and Relations). М. 1963.
[33] A. Ujomov: Gesetz und innere Relation. „Gesetzbegriff". Berlin 1960.
[34] Реология. Теория и приложения. Под редакцией Ф. Эйриха (Rheology, Theory and Applications. F. Eyrich, Ed.). М. 1962.
[35] О. Ф. Погорелов: Системный подход к аиалнзу процесса генезиса (O. F. Pogorelov: System Approach to the Analysis of Genesis Problem). In: Философьскі проблеми сучасного природознавства. Выпуск 27. Київ 1972.

ENTROPY AND INFORMATION IN THE LIFE SCIENCES

II

INFORMATION AND BIOLOGICAL THERMODYNAMICS

K. S. Trincher

1.

The first mental study in which mention is made of information as a parameter with a physical significance is Maxwell's concept of a demon capable of differentiating molecules according to their speed. Maxwell's demon is a mechanism endowed with an intellect; it performs rational functions not requiring energy: the opening and closing of a microscopic door in the wall dividing a gas cylinder into two halves. The intellectual mechanism passes "hot" molecules into one half of the cylinder, and "cold" molecules into the other; as a result of this sorting each half of the cylinder acquires a different temperature:

$$T_1 > T_{ex} > T_2,$$

where T_{ex} is the original temperature which is equal to the temperature of the external medium from which the cylinder is separated by an adiabatic shell, T_1 is the temperature of the "hot" half, T_2 is the temperature of the "cold" half. The temperature gradient that has become established inside of our mental system which includes an intelligent being, makes it possible to connect up to the system a heat engine that can operate on the bound energy of the gas in the "hot" half of the cylinder. The work of the heat engine is determined according to the equation:

$$-\int_{T_1}^{T_2} S\,\mathrm{d}T = A + q,$$

where S is the entropy of the gas, A is work, q is the so-called uncompensated heat which inevitably accompanies any irreversible work process [9]. Let us assume that the work of our heat engine is "quasi-ideal", i.e. practically frictionless. In this case the uncompensated heat, measured in the same units, is much less than the work and it can be disregarded:

$$A \gg q \approx 0.$$

From here we obtain the expression for the work of the quasi-ideal heat engine:

$$S\,.\,(T_1 - T_2) = A^*.$$

In our mental study, all the physical objects and parameters used (external medium, cylinder, gas, entropy, temperature, heat engine, and work) are of macroscopic dimensions, while the intellectual mechanism (Maxwell's demon, which distinguishes molecules according to their speed and passes them through a microscopic aperture) functions as the connecting link between micro-and macro-dimensions. Due to this ability of the demon, the kinetic energy of the chaotically moving gas molecules is transformed into the work of the heat engine. We can now say that this function, the role of connecting link between the micro-and macro-phenomena of nature, can be considered a characteristic property of information in general. We find this unique property of information in the response of an organism to signals from the external medium. The signal perceived by the sensory cell is of a microscopic nature, while the response of the organism is a macroscopic work process.

When the work of the quasi-ideal heat engine is completed, the tempeature in both halves of the cylinder is the same, that is, it equals T_2. Now let us remove the adiabatic shell and create the conditions in which the gas is heated by an influx of heat from the external medium. The outer environment is assumed to be infinitely large compared to the size of the cylinder, and by transferring heat to the gas its own temperature does not drop. After the gas regains its original temperature T_{ex}, we "clothe" the cylinder in the adiabatic shell and again take advantage of the ability of the intellectual mechanism, that is Maxwell's demon. As soon as a different temperature is reached in the two halves of the cylinder due to sorting of the molecules, we again connect up the quasi-ideal heat engine and the whole work process taking place at the expense of the bound energy of the gas in the hot half of the cylinder is repeated. Thus, despite the second principle of thermodynamics, Maxwell's demon makes it possible for us to obtain a periodically operating machine utilizing the heat of the external medium to perform work.

Many years passed before Maxwell's demon was deposed. The immaterial intellect, capable of purposefully employing the micro-information it received, was nevertheless incapable of overcoming the obstacle created by the second law of thermodynamics. This obstacle states: even an intelligent being capable of receiving information on the properties of individual molecules in the vast multitude of molecules in a gas medium cannot, with the help of clever mechanisms, employ the bound energy of the gas for useful cyclicly repetitive work. In 1929 Szilard [16] proved that the receipt of information from molecules in the form of light signals, for example, is connected with the expenditure of exactly the same amount of free energy as can be obtained from the hot half of the cylinder and utilized for the work of the heat engine. The mental creation of the great physicist became an odious figment of the imagination, and his demon was reduced to naught before the might of the second principle of thermodynamics.

However, Szilard permitted himself a certain amount of arbitrariness when he ascribed to Maxwell's demon properties that are by no means mandatory. Actually, if an intelligent mechanism actively processes information—differentiates "hot" and "cold" molecules with the help of sensory heat receptors, and passes molecules through a microscopic aperture – then this demon, functioning as a preliminary apparatus of the heat engine, is precisely the macroscopic system whose actions are all subordinated to the second principle of thermodynamics. However, let us imagine Maxwell's demon as something other than a macroscopic system separately performing the functions of information (distinguishing individual molecules) and mechanical functions (opening and closing microscopic doors). We can imagine Maxwell's demon in the material shape of a microscopic elementary machine capable of combining the receipt of information – signals of a microscopic nature – and the performance of internal structural work within the elementary machine. The internal work is not performed in relation to some kind of "external" thermodynamic force acting from an external medium; it is performed in relation to an internal force acting inside the elementary machine. Such elementary machines capable of using heat as a source of energy and combining internal work with the information process actually exist. They are the working biological macromolecules: "the hemoglobins" [20].

2.

Before passing to a description of the work of the elementary machine – the hemoglobin molecule which biochemists call an "honorary enzyme" – and then undertaking a description of the simplest "biomachine" – the metabolizing cell in the condition of basal metabolism – it is necessary to point out the different contents of the concept of information, considered above as a parameter with a physical significance. As already indicated, one of the indispensible properties of information is to be the connecting link between the micro- and macro-phenomena of nature. The micro-phenomena embrace the gas molecules, the oxygen molecules and all possible signals from the external medium. The corresponding macro-phenomena are the work of the heat engine, the internal (structural) work of the hemoglobin molecules, and the response of the sensory cell to the stimuli. It is obvious that information is a parameter with a physical significance that does not, however, possess dimensionality. This gives rise to the following question: How does a dimensionless parameter, information, realize the connection between the micro- and macro-phenomena which have dimensions. To solve this problem let us examine the functional position of information in the following scheme of the three processes:

(A)

Kinetic energy of gas molecules } → { Information on velocity of gas molecules received by Maxwell's demon
⇕
→ { Work of heat engine

(B)

Heat entering hemoglobin molecule from external medium } → { Information of hemoglobin macromolecule on release of oxygen
⇕
→ { Structural work of hemoglobin molecule

(C)

Signal received in sensory cell from external medium → { Information received by sensory cell in form of stimulus

⇕

→ { Response of sensory cell as external work

In scheme (A) information and the work of the heat engine are represented in the most abstract form: the information is the spiritual content of a purely mental being – Maxwell's demon, and the heat engine figures as an aimlessly operating mechanism. However, scheme (A) can serve to illustrate any concrete work process performed by a purposefully functioning machine. The design of a purposefully operating machine is the structural information imparted to it by the human intellect. Schemes (B) and (C) show that the purposeful structure of the hemoglobin molecule and of the living cell, and the work performed by them represent inter-connected phenomena.

In conformity with these schemes illustrating "work" at all known levels – technical, macromolecular and biological – information is presented as the accompanying link of work, indicating the objective of the work process. Information as a physical and dimensionless parameter exists only in connection with a work process that has a definite objective. Information must be connected with purposefully operating systems such as the man-made machine, the working macromolecule, and the metabolizing cell; without this connection information does not exist at all.

In reality, the processes that take place in non-living nature and not connected with the organic world do not have the nature of purposeful work and do not contain any information indicating that the process taking place has any objective whatever. In non-living nature no objective exists, and all events occurring in it are described by physical laws of a deterministic or probability nature.

Thus, we can state that the properties of biological information are:

(a) the function of connecting link between micro- and macrophenomena, that is between the signal and the objective;

(b) the role of accompanying link of purposeful work.

Let us picture these properties of biological information in the form of the following scheme:

signal → information → objective

⇕

work

↗ ↑ ↖

machine biological macromolecule cell

Having in this way defined the biological essence of information as the accompanying link of any purposeful work at the level of the man-made machine, the biological macromolecule and the functioning cell, we must not overlook the definitions of information elaborated by mathematicians and physicists in their study of the measurement process. We shall show that the biological, mathematical and physico-theoretical definitions of information represent different aspects of this concept, which are not contradictory.

From the purely formal, mathematical position information was connected with the measurement process which is the source of information; in this aspect information was quantitatively expressed by a function resembling the Boltzmann entropy [17; 22]:

$$H(x) = -\sum_{i} p(i) \, . \, \lg_2 p(i),$$

where $H(i)$ is the quantity of information that can be obtained from a certain group of events x, classified according to various categories i; the symbol $p(i)$ represents probability of the specific event.

According to Brillouin [5], the quantity of information ΔJ that can be obtained by measurement equals

$$\Delta J = k \, . \ln \frac{P_0}{P_1},$$

where k is the Boltzmann constant, P_0 is the number of equally probable cases before measurement, P_1 is the number of cases left after measurement. An important concretization of the concept of information in the physical aspect was obtained by Brillouin on the basis of the connection of the value ΔJ with the value ΔS, the entropy increase due to the measurement process. Thus we obtain the following relationship:

$$\Delta S \geqq \Delta J.$$

Due to the irreversibility of the measurement process an entropy increase is observed which in the quantitative aspect is greater than the increase in information. Since the measurement process involves consumption of energy, the entropy increase is determined by the quantity of heat into which the energy is transformed while the measurement is made at temperature T. From here we obtain the connection between the energy dissipated in the form of heat due to the measurement process and the "useful energy" accumulated as the product of the temperature and the quantity of information obtained:

$$T \, \Delta S \geqq T \, \Delta J = Tk \, . \ln \frac{P_0}{P_1}.$$

This relationship leads to the extremely important conclusion made by Brillouin: the more accurate a measurement is, that is the smaller the value P_1 is in relation to the value P_0, the greater the entropy increase. From here it follows that if $P_1 = 0$, that is if measurement were absolutely precise or if the result of the measurement were absolutely definite, the entropy arising in the process of measurement would grow ad infinitum. Thus, the physical definition of information, like the mathematical definition, reflects the various aspects of this concept and does not contradict the biological definition which is connected with the performance of internal work.

3.

Let us now proceed to an investigation of the elementary machine – the biological macromolecule hemoglobin – and show, by means of this simplest biological object performing stereotype purposeful work, that there exist two parallel processes having the nature of information and energy. Let us symbolically note these processes taking place in the hemoglobin molecule:

(A) ... signal → information → objective,
(B) ... heat → work → energy,

where (A) represents the information (dimensionless) chain and (B) represents the energy (dimensioned) chain.

The signal, penetrating the biological macromolecule, starts off the information process—the rearrangement of the structure of the hemoglobin molecules. The end of the information process, that is the end of the structural rearrangement of the macromolecules, represents attainment of the objective which is a new and more highly ordered structure of the biological macromolecules. Attainment of a more highly ordered structure signifies the following: the biological macromolecule has increased its volume, and the ordering of its structure, consequently, reaches greater dimensions. The energy process takes place parallel with the information process: the biological macromolecule absorbs the heat and transforms it into structural work. When the structural work is completed, structural energy is accumulated in the hemoglobin macromolecule. Completion of the structural work coincides with completion of the information process and represents the objective attained: the state in which the hemoglobin contains the maximum amount of energy, is characterized by a maximum volume and a maximum ordering of the structure.

After completion of the information process and along with it the work process, that is after attaining the objective which represents the extremal condition characterized by the three maximums—energy, ordering of the structure and volume—the inverse process begins. The return of the hemoglobin molecules to the initial position also takes place as two parallel processes (A′) and (B′) which are opposite to the direction of (A) and (B). The inverse process takes place without the intermediate links of information and work. The return of the hemoglobin molecule to the initial position characterized by the three minimums—energy, volume and ordering of the structure—can be illustrated as follows:

(A′) ... objective → minus signal
(B′) ... energy → minus heat,

where (A′) is the dimensionless chain representing the loss of structural information (loss of ordering of the structure), and (B′) is the dimensioned chain representing the transformation of the structural energy into the heat released from the hemoglobin molecule.

In the information chain (A) the positive signal is connected with the ordered structure; in the inverse chain (A′) the ordered structure is connected with the minus signal. A similar inversion takes place in the (B) and (B′) chains. The structural energy that accumulates in the biological macromolecule as a result of the structural work changes into minus heat, that is into the heat released from the biological macromolecule. Simultaneously with the appearance of the minus signal and the minus heat, there takes place a reduction in the volume of the hemoglobin molecule, the elementary machine returns to its initial position and the work cycle, conjugate with the information process, begins all over again.

On the basis of the "information-thermodynamic" scheme presented here, let us now examine the structure-functional alterations of the hemoglobin molecule in terms of concrete, experimentally established data. The take up and release of oxygen correspond to the minus and plus signals of the information chain. When oxygenation takes place there is a decrease in the volume of the hemoglobin molecule $-\Delta V$, and during de-oxygenation there is an increase in volume $+\Delta V$. Simultaneously with the structural changes in the biological macromolecule energy changes take place: Oxygenation is accompanied by heat liberation $-\Delta Q$, and de-oxygenation, by heat absorption $+\Delta Q$. Thus in oxygenation and de-oxygenation the biological macromolecule is subject to a triple process: chemical, energy and structural. Schematically this can be noted as follows:

- chemical reaction: intake and release of oxygen,
- energy transformation: intake and release of heat,
- structural alteration: increase and decrease in volume.

Let us now describe these three parallel processes in the form of equations containing energy and geometric symbols as well as the chemical symbols HbO_2, Hb and O_2. Oxygenation is expressed as follows:

$$\mathrm{Hb} + \mathrm{O_2} - \Delta Q - \Delta V = \mathrm{HbO_2}.$$

The opposite process, de-oxygenation, is expressed as follows:

$$\mathrm{HbO_2} + \Delta Q + \Delta V = \mathrm{Hb} + \mathrm{O_2}.$$

Let us now determine what internal changes take place in the hemoglobin molecule during oxygenation and de-oxygenation. For this purpose the three parallel processes—chemical reaction, transformation of energy and structural alteration—shall be noted separately:

$$\text{oxygenation}\quad \begin{cases} \mathrm{Hb} + \mathrm{O_2} = \mathrm{HbO_2} & \ldots \text{ oxygen intake,} \\ U_{\mathrm{Hb}} - \Delta Q = U_{\mathrm{HbO_2}} & \ldots \text{ heat release,} \\ V_{\mathrm{Hb}} - \Delta V = V_{\mathrm{HbO_2}} & \ldots \text{ volume decrease,} \end{cases}$$

$$\text{de-oxygenation}\quad \begin{cases} \mathrm{HbO_2} - \mathrm{O_2} = \mathrm{Hb} & \ldots \text{ oxygen release,} \\ U_{\mathrm{HbO_2}} + \Delta Q = U_{\mathrm{Hb}} & \ldots \text{ heat intake,} \\ V_{\mathrm{HbO_2}} + \Delta V = V_{\mathrm{Hb}} & \ldots \text{ volume increase,} \end{cases}$$

where U_{Hb} and V_{Hb} are the internal energy and the volume of the hemoglobin molecule, and $U_{\mathrm{HbO_2}}$ and $V_{\mathrm{HbO_2}}$ are the internal energy and the volume of the oxyhemoglobin.

To proceed with the analysis of the structure-energy changes in the hemoglobin molecule it will be necessary to introduce two thermodynamic notions: the structural energy of the hemoglobin molecule ΔU_{str}, and the structural (internal) work performed by the hemoglobin

A_{str}. The structural energy is determined by two equations (a) and (b). In equation (a) the structural energy is determined as the difference of the internal energies of molecules Hb and HbO_2:

$$\Delta U_{str} = U_{Hb} - U_{HbO_2}, \tag{a}$$

and in equation (b) the structural energy is expressed as the result of the transformation of heat through structural (internal) work:

$$\Delta Q = A_{str} = \Delta U_{str}. \tag{b}$$

This new thermodynamic value, structural work, that is the process of transformation of heat into structural energy, first appears in the working biological macromolecule: hemoglobin. The structural work takes place against the intramolecular pressure of the hemoglobin molecule and leads to an increase in its volume. Thus the structural work is determined by the following equation:

$$A_{str} = N_0 p_{Hb} \Delta V,$$

where N_0 is the Avogadro number, p_{Hb} is the intra-molecular pressure of the hemoglobin molecule. From here we obtain

$$p_{Hb} = \frac{A_{str}}{N_0 \Delta V}.$$

Taking into account the equality between the structural work and the amount of heat absorbed by the hemoglobin molecule, we have

$$p_{Hb} = \frac{\Delta Q}{N_0 \Delta V}.$$

We have obtained an equation which permits calculation of the intra-molecular pressure of the biological macromolecule from experimentally established data – the heat of the reaction and the decrease in the volume of the hemoglobin molecule in oxygenation.

Numerically, the heat discharged when one mole of oxygen and one mole of hemoglobin are combined is equal to [1; 2]:

$$\Delta Q = 14 \text{ Cal . mole}^{-1}.$$

Let us simplify certain assumptions to obtain the numerical value for the difference in the volume of the hemoglobin molecule and the oxyhemoglobin. We shall assume that molecules Hb and HbO_2 are spheres with radii r_{Hb} and r_{HbO_2}.

Since $r_{Hb} \simeq 25$ Å, we obtain the following value for the volume of the hemoglobin molecule:

$$V_{Hb} = \frac{4}{3} r_{Hb}^3 \pi = 6 . 10^{-20} \text{ cm}^3.$$

We shall determine the radius of the oxyhemoglobin molecule using the data obtained by X-ray analysis of its structure. The linear reduction of the HbO_2 molecule, determined from the displacement of the iron atoms by a comparison of the X-rays of the structure of the Hb- and HbO_2 molecules, is approximately 7 Å [13; 14]. From this it follows that the radius of molecule HbO_2 is about 3 Å smaller than the radius of molecule Hb. By accepting the value $r_{HbO_2} = 22$ Å we obtain the following value for the volume of the oxyhemoglobin molecule:

$$V_{HbO_2} = \frac{4}{3} r^3_{HbO_2} \cdot \pi \simeq 4{,}3 \,.\, 10^{-20}\ \text{cm}^3.$$

In this way, by making use of the numerical value for the volume of hemoglobin and oxy-hemoglobin, we shall determine their difference:

$$\Delta V = V_{Hb} - V_{HbO_2} = 1{,}7 \,.\, 10^{-20}\ \text{cm}^3$$

Using the values $\Delta Q = 1{,}4 \,.\, 10^4$ cal . mol^{-1}, $N_0 \simeq 6 \,.\, 10^{23}$, $\Delta V = 1{,}7 \,.\, 10^{-20}$ cm^3, 1 cal = = 41 cm^3. at., we obtain

$$p_{Hb} = \frac{\Delta Q \,.\, 41}{N_0 \,.\, \Delta V} \cong \frac{1{,}4 \,.\, 10^4 \,.\, 41}{6 \,.\, 10^{23} \,.\, 1{,}7 \,.\, 10^{-20}} \cong 57\ \text{at.}$$

Thus we have determined the numerical value of a new thermodynamic parameter – intra-, molecular pressure – characteristic of a working biological macromolecule. As can be seen, the internal pressure of the hemoglobin molecule is very small. The low internal pressure of molecule Hb indicates that it is related to the bond between the easily moving subunits of the hemoglobin molecule. In reality, the conformational changes in the hemoglobin molecule, the decrease in volume when oxygen is bound and the increase in volume when oxygen is released, takes place only in the intact molecule, that is when all the four subunits constituting the hemoglobin molecule are simultaneously present. In the individual subunit, in the α- or β-chain, there is no change of volume during binding or release of oxygen. A change in volume during oxygenation and de-oxygenation can be found only in the intact hemoglobin molecule. From this we reach the important conclusion that the change in the volume of the hemoglobin molecule as a whole is a function of the mobility of its subunits, the two α- and the two β-chains, relative to each other.

The subunits of the hemoglobin molecule possess a high mobility; this is indicated by the low value of the intramolecular pressure of the hemoglobin molecule. Actually, by comparing the internal pressure of the hemoglobin molecule with that of the fluids we see that p_{Hb} is two orders less than the internal pressure of weakly associated highly fugative fluids, which ranges from $3 \,.\, 10^3$ to $6 \,.\, 10^3$ at. We can, therefore, speak of the "ultra-mobility" of the α- and β-chains in the hemoglobin molecule. This intrinsic structural-kinetic property – the ultra-mobility of the subunit – explains the hemoglobin molecule's ability to absorb heat and transform it into structural work without loss.

Let us now replace the above equations for oxygenation and de-oxygenation containing chemical, energy and geometric symbols, by equations representing the "information-thermo-dynamic changes" of the hemoglobin molecule acting as a purposefully working elementary machine. Let us again examine the scheme of the parallel information and energy processes corresponding to de-oxygenation:

(A) ... signal → information → objective,

(B) ... heat → work → energy.

By applying the notion of structural energy and the expression for structural work we obtain the following equation for de-oxygenation:

$$HbO_2 - O_2 + N_0 p_{Hb} \Delta V = Hb + \Delta U_{str}.$$

The meaning of the symbols is:

$HbO_2 - O_2$ – signal—exit of oxygen molecule from the hemoglobin molecule;

$N_0 p_{Hb} \Delta V$ – internal work of hemoglobin molecule which simultaneously represents the information process. Work is performed due to heat absorbtion by hemoglobin;

Hb – objective of the information-thermodynamic process, that is ordering of the structure, which is expressed by an increase in the volume of molecule Hb as compared to HbO_2;

ΔU_{str} – structural energy accumulated in the hemoglobin molecule as a result of internal work.

The inverse process, oxygenation, which takes place without intermediate information and work links, is represented by the following scheme:

(A′) ... objective → minus signal,
(B′) ... energy → minus heat.

The equation for oxygenation corresponding to these two parallel inverse processes appears as:

$$Hb + O_2 - \Delta U_{str} - N_0 p_0 \Delta V = HbO_2 + \overrightarrow{\Delta Q}.$$

The meaning of the symbols here is:

Hb – disappearance of the objective;

O_2 – minus signal—penetration of oxygen into the hemoglobin molecule;

$-\Delta U_{str}$ – structural energy that changes into heat;

$-N_0 p_0 \Delta V$ – decrease in volume taking place when internal pressure has a zero value and expresses decreased ordering of structure;

$\overrightarrow{\Delta Q}$ – minus heat, i.e. the heat into which the structural energy is transformed and which is released from molecule Hb.

We can now formulate our conclusions as follows: When "expansion" in the volume of the hemoglobin molecule takes place, the heat absorbed is transformed through internal-structural work into structural energy; when, "compression" of the hemoglobin molecule takes place, no work is performed and the structural energy that has been stored in the expanded volume is converted directly into the heat released from the hemoglobin at the time the volume is decreased. We have in this way discovered new structural and thermodynamic properties characteristic of the elementary machine—the biological macromolecule, hemoglobin:

(a) the process of conversion of heat through internal work into structural energy takes place without entropy; this process is therefore ideally equilibrated;

(b) the subunits of the hemoglobin molecule possess "ultra-mobility"; they move against the internal pressure of the hemoglobin molecule, which is extremely low.

Until recently the change in the volume of the hemoglobin molecule could not be explained [11; 8; 13; 23]. Our information-thermodynamic analysis has solved this problem and overcome the barrier imposed by the second fundamental principle of thermodynamics. The biological macromolecule hemoglobin performs internal, structural work at the expense of the heat absorbed, and as a result of this work which takes place without entropy, the heat is completely converted into structural energy.

Let us here recapitulate the thermodynamic equation for this processess:

$$\Delta Q = A_{\text{str}} = \Delta U_{\text{str}}.$$

As distinct from the biological macromolecule hemoglobin, the mechanical machine that works at the expense of heat or free energy, performs external work which, being irreversible, is accompanied by entropy. The thermodynamic equation that describes the work process of a mechanical machine is as follows:

$$\left.\begin{matrix}\Delta Q\\ \text{or}\\ \Delta F\end{matrix}\right\} = A + T\Delta S_{\text{uncomp}},$$

where F is free energy, A is external work, T is absolute temperature, S_{uncomp} is the so-called uncompensated entropy that unavoidably arises during the performance of external work.

Between the mechanical and elementary machines there is, consequently, a fundamental thermodynamic difference: the external work of the mechanical machine is performed in conformity with the second principle of thermodynamics, the external work is accompanied by the production of entropy, while in hemoglobin the internal (structural) work of this biological macromolecule takes place without the production of entropy.

Let us now proceed to an investigation of the elementary machine of a higher order – the living cell which continuously performs the internal work of basal metabolism – the work of preserving its structure.

4.

The living cell in a state of basal metabolism does not perform any external work, but expends energy only on internal work to preserve its structure. A thermodynamic model of a cell in the state of basal metabolism is shown in Fig. 1. Ordering of the cell structure is represented in Fig. 1(a) in the form of alternating "elementary layers" distinctly separated from each other by their surfaces. Fig. 1(b) presents one elementary layer consisting of sub-layers P and W: the P-layer represents protein, and the W-layer, intracellular water; the relationship between P- and W-layers is approximately – 1 : 4. This elementary two-level layer of protein and water represents the structural-energy unit of living matter.

In the elementary layers of the metabolizing cell a continuous work cycle takes place at the expense of the free energy of nutritive matter which enters the cell. During performance of internal work an equivalent amount of heat is liberated from the cell. The relation between the speed at which free energy is consumed $(dF/dt)_N$, the speed of the performance of the internal work of preserving the layer structure $(dA/dt)_{str}$ and the speed of liberation of an equivalent amount of so-called compensated heat $(dQ/dt)_{comp}$ is schematically shown in Fig. 1(a).

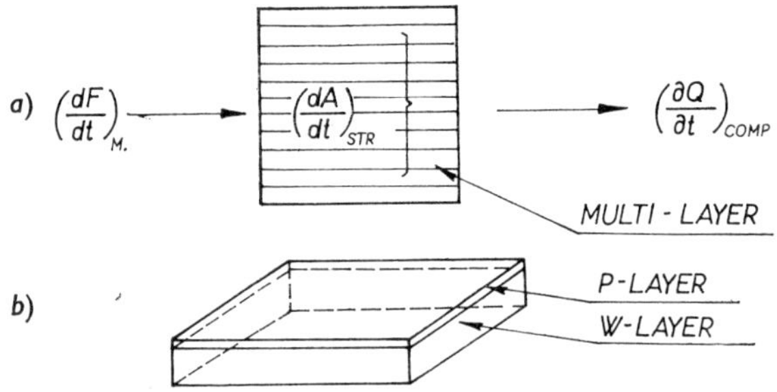

Fig. 1. Thermodynamic model of the cell.

By means of the information thermodynamic analysis, let us now study the structural-energy changes in the elementary layer of the cell in a state of basal metabolism. Similarly to the work of the elementary machine—the hemoglobin molecule—in the elementary layer we have two parallel processes that take place simultaneously:

(A) signal → information → objective,
(B) free energy → structural work → structural energy.

Since the cell is in a state of basal metabolism, signals from the outside do not penetrate into it, but a signal originates periodically within the cell itself in the form of disordered structure. The objective of the information process (A) represents ordered structure:

(A) disordered structure → information → ordered structure.

The inverse process taking place without the information link and without structural work appears as follows:

(A′) ordered structure → disordered structure,
(B′) structural energy → heat.

Let us now pass to the thermodynamic analysis of the work cycle. Fig. 2 is a diagram of the structural and thermodynamic changes that take place in the elementary layer in the course of a single work cycle. In the left half of the diagram the protein structure and the water are represented in the most disordered condition, while in the right half they are most ordered. The protein molecule is represented at the left in the form of a random coil, and the intracellular water—in the form of fluid. These conditions of greatest disordering of the structure are characterized by the following expression of the internal energy of the protein and water:

$$U_{prot} = T_{cell}S_{prot} \quad \text{and} \quad U_w = T_{cell}S_w,$$

where T_{cell} is the temperature of the metabolizing cell, S_{prot} and S_w are the entropies of the protein and the water.

Although the protein molecule in the form of a random coil and the intracellular water in the fluid state possess thermal stability, their state in the elementary layer is unstable. This state of the two layers is possible but not probable, and in time the protein and water molecules should become mixed. However, in the living cell this does not happen since the stratified structure of the protein and fluid water is preserved as a result of internal work.

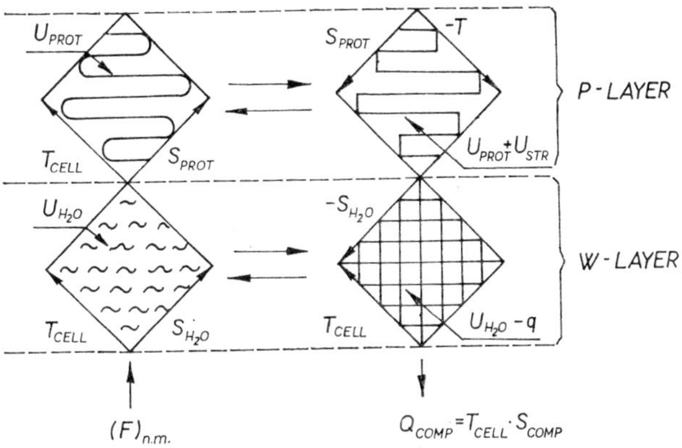

Fig. 2. The structure-energy changes in an elementary layer of the cell.

Chemical energy from the outer medium in the form of glucose molecules, for example, penetrates into the intracellular water. The agent of the chemical energy passes through the water layer and comes into contact with the active center of the protein-enzyme molecule. In extracting the chemical energy from the molecular substrate and transforming the chemical energy into structural energy, the protein molecule undergoes a structure-energy change: from the state of a random coil, the protein molecule changes into a superordered, structurally stressed state. Simultaneously with the structure-energy change in the protein molecule, a structure-energy change takes place in the intracellular water. Under the forces of the molecular field of the structurally-stressed protein, a phase change of water takes place: "fluid → crystal". The phase change of intracellular water is accompanied by heat liberation from the elementary layer to the extracellular medium. The super-ordering of the state of the protein molecule and the ordering of the crystal state of the intracellular water is shown at the right in Fig. 2. In a state of structural stress, the energy of the protein molecule is equal to

$$U_{prot} + \Delta U_{str},$$

and the energy of the intracellular water in the crystal state is equal to

$$U_w - \Delta q,$$

where Δq is heat liberated from the water during the "fluid → crystal" phase change. Let us consider the internal energies of the protein and the water as zero levels of energy. Then the state of the elementary layer, consisting of super-ordered protein molecules and crystallized water, is characterized by polarization of energy. In the "upper" pole, surplus energy is created due to absorption of the structural energy of the protein molecule, and in the "lower" pole, an energy deficit, a "quasi-energetic vacuum" appears due to the liberation of heat in the phase change of intracellular water:

$$\begin{Bmatrix} +\Delta U_{str} \\ -\Delta q \end{Bmatrix}.$$

The structure of the elementary layer in a state of energy polarization is thermally labile, or thermally unstable in time. For this reason the structural energy of the super-ordered protein molecule passes to the crystallized water, becoming transformed into the internal energy of the water. The protein molecule, deprived of its structural energy changes from a high ordered state of structural tension to the random coil state, while the intracellular water, having received energy, performs the "crystal → fluid" phase change. This completes the work cycle. A portion of free energy again penetrates the fluid water and the work cycle, being of an auto-oscillating nature, begins again.

We have designated the energy deficit of intracellular water in the crystalline state by the negative thermodynamic term "negative heat." Let us now determine the state of crystallized water, characterized by an energy deficit, by means of negative entropy; this term expresses the ordering of the structure of water, according to the equation

$$d(-S)_w = \frac{\partial(-q)}{T_{cell}}.$$

The function of this state—the negative entropy—expresses the ordering of the structure of the water due to liberation of heat during the "fluid → crystal" phase change. Consequently we have two thermodynamic negative terms $-S_w$ and $-q$ which characterize the ordered state of the water within the cell.

Let us now show that two negative thermodynamic terms are also necessary to characterize the ordered state of the protein molecule. During the change from the random coil state to the super-ordered state of the structural tension, the protein molecule performs structural (internal) work according to the equation:

$$dF_N = dA_{str} = dU_{str}.$$

Let us express the work of ordering the protein structure through a factor of intensity, the negative absolute temperature, and through the volume factor, the negative entropy of the protein molecule:

$$dA_{str} = (-T)\, d(-S)_{prot}.$$

From the latter two equations we obtain the following expression for the function of the state where the protein molecule has a super-ordered structure:

$$d(-S)_{prot} = \frac{dU_{str}}{-T}.$$

The negative absolute temperature has been introduced in thermodynamics to characterize the system in a particular state of equilibrium. At a positive absolute temperature in a state of equilibrium the system has a minimum energy, while at a negative absolute temperature it has a maximum energy. The maximization of the energy at negative absolute temperature corresponds to the super-ordered state of the system [15; 7; 10].

As a result, we can describe the metabolism of the elementary cell layer in a state of basal metabolism in the form of a scheme showing the irreversibility of the process of energy transformation and the reversibility of the structure-energy changes of protein and water during the work cycle:

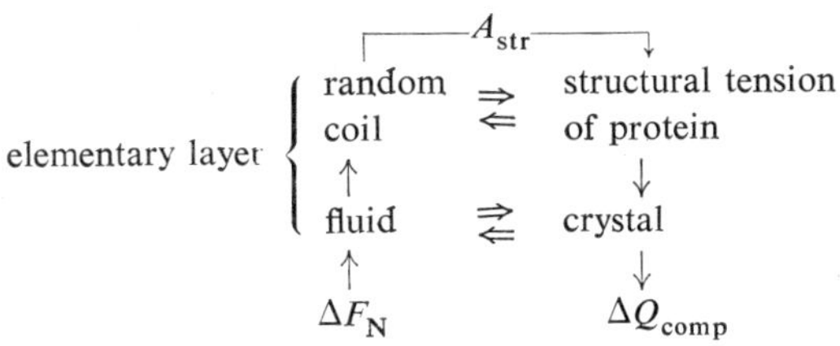

The irreversible process of energy transformation in the living cell is described by the following thermodynamic equation:

$$\Delta F_N = A_{str} = \Delta Q_{comp}.$$

The cell consumes free (chemical) energy of nutrative matter, performs structural (internal) work without an entropy arising and liberates an equivalent amount of heat. We see that in the thermodynamic as well as in the information sense the process of energy transformation in the cell differs fundamentally from the processes of energy transformation in the mechanical machine. In the man-made machine, the structure, that is the programme embodied in its design, is a prerequisite of the work. The entropy arising due to the work is the cause of the destruction of the machine. In the living cell work, which is of the nature of internal work, is a prerequisite for the preservation of the structure. This work is performed without entropy and its termination inevitably leads to destruction of the cell structure. This brings us to the following fundamental thermodynamic conclusion [18]:

"The structure of the cell is thermally labile at its vital temperature; the preservation of the cell structure requires continuous internal work of a cyclical nature, each cycle being performed with the consumption of a quantum of free energy and liberation of an equivalent amount of heat."

From the discreteness of the work cycle in time it follows that the internal work of the cell and the heat liberated from it have the nature of quanta:

$$A_{str} = \frac{h}{\tau_{cell}} \quad \text{and} \quad \Delta Q_{comp} = k \,.\, \Delta T,$$

where h is Planck's constant, τ_{cell} is the time of the work cycle, k is the Boltzmann constant, ΔT is the difference between the temperature of the isothermically working cell and the external medium to which the heat is liberated.

The equality of internal work and heat leads to the formulation of "the fundamental law of cell metabolism"; this law connects the continual values characteristic of the phenomena of life—time and temperature—with the discrete values, the universal physical constants h and k, characteristic of the quantization phenomena of non-living nature:

$$\tau_{\text{cell}} \,.\, \Delta T = \frac{h}{k}\,.$$

The fundamental law of cell metabolism was verified experimentally by conducting a single impulse through a nerve fibre [19; 21].

We investigated the stereotype work of a cell performing only one function—the function of existence, that is the preservation of its structure which is thermally labile at vital temperature. Let us now apply our information-thermodynamic analysis to the most complex phenomenon of life—the ontogeny of a living organism from the moment of fertilization of the egg cell to the mature stage.

5.

The ontogeny of the animal organism, consisting of approximately $\sim 10^{10}$ cells, represents a vast process of ordering which takes place in three aspects: time, space and matter. During ontogenesis, from the moment of fertilization of the egg cell, a certain "time counter" mechanism (t-function) operates which determines the sequence of the stages of development. Simultaneously with the time counter, a "space (geometrical) ordering (g-function) mechanism" operates. At the end of the growth phase all the $\sim 10^{10}$ differentiated cells that have originated from a single fertilized egg cell occupy a definite position in the space of the organism. Both functions of time-space ordering are combined with the functional ordering of the mass of the differentiated cell (m-function). At the end of its development, the differentiated cell performs a certain specific function or number of specific functions connected with the ordering of its structure. Thus the fertilized egg cell, representing the "origin of development," in other words "the beginning of the information-energy process," has a triple relationship with the objective of development, that is with the nature organism. We call this triple relationship, which corresponds to ordering in time and space, and to the specific functions of the cells, "trajectories of development." By using the symbols t, g and m we shall present the information-energy process in the ontogeny of the animal organism in the form of the following symbolic record:

$$O \left\{ \begin{array}{l} \xrightarrow{t} \\ \xrightarrow{g} \\ \xrightarrow{m} \end{array} \right. \sum_{1}^{10^{10}} (\text{cell})_i ,$$

where O is the origin of development, that is the beginning of the information-energy process—the fertilized egg cell, $\sum_{1}^{10^{10}} (\text{cell})_i$ is the objective of development, that is approximately $\sim 10^{10}$

differentiated cells $(\text{cell})_i$, each of which is connected with its own individual trajectory of development through t, g and m-functions.

Let us combine two of the trajectories of development, functions t and g, in a single common function of ordering in time and space: $\omega(t, g)$. Then, let us substitute the function of specific activity connected with the structure of the differentiated cell through the function of growth of the mass of the animal organism: $m \equiv m(t)$. We now introduce a new function, the specific function of ordering, which represents the relationship of the general function of ordering $\omega(t, g)$ and the function of growth of mass of living matter:

$$f(t) = \frac{\omega(t, g)}{m(t)} .$$

Let us now determine the energy nature of the specific function of ordering by the following equation:

$$f(t) = \frac{1}{m} \cdot \frac{dQ}{dt},$$

where dQ/dt is the speed of heat generation of the developing organism, m is the mass of the organism from which the heat is liberated.

Let us express the connection between the energy content of the specific function of ordering and the informational content in the form of the following equation:

$$\frac{\omega(t, g)}{m(t)} = \frac{1}{m} \frac{dQ}{dt} = f(t).$$

Thus the specific function of ordering $f(t)$ is of a dual nature:

informational, which is the realization of genetic information during ontogenesis;

energetic, which is the consumption of free energy and the liberation of an equivalent amount of heat during ontogenesis.

Thus we obtain an equation permitting us to find the dependence of the specific heat generation of the developing organism on the time of ontogenesis:

$$\frac{1}{m} \frac{dQ}{dt} = \text{function of time.}$$

This function becomes known when $\omega(t, g)$ and $m(t)$ are known as functions of time.

Let us now proceed to a definition of these functions and establish the theoretical dependence of $(1/m)(dQ/dt)$ on time. It is obvious that the speed of ordering of mass in the developing organism decreases as the time interval from the moment of fertilization of the egg cell becomes greater:

$$\frac{d\omega(t, g)}{dt} = K \cdot \frac{1}{t},$$

where K is the factor of proportionality. From here

$$\omega(t, g) = K \ln t \,\Big|_{t_{in}}^{t_{fin}},$$

where t_{in} is the initial moment of ontogenesis, and t_{fin} is the final moment coinciding with the end of the growth of the organism.

On the basis of thermodynamic conceptions, Ludwig von Bertalanffy evolved the equation of the growth of the animal organism [3; 4]:

$$m(t) = \{(m_{fin})^{1/3} - [(m_{fin})^{1/3} - (m_{in})^{1/3}]\, e^{-K't/3}\}^3 \equiv f(B),$$

where m_{fin} and m_{in} are mass of the animal organism at moments of time t_{fin} and t_{in}; $f(B)$ is the Bertalanffy function which expresses the S-shaped character of the dependence of mass of the growing organism on time.

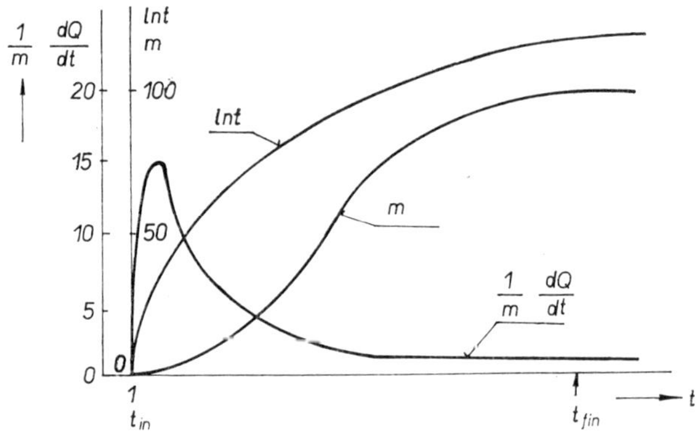

Fig. 3. The theoretical dependence of the function on time of ontogenetic development.

By applying the expression for the function of ordering and the Bertalanffy function we obtain the following equation connecting the specific heat generation of the organism with the process of realization of genetic information during ontogenesis:

$$\frac{1}{m}\frac{dQ}{dt} = K \cdot \frac{\ln t}{f(B)}\Bigg|_{t_{in}}^{t_{fin}}.$$

We have called this equation the thermodynamic principle of ontogenetic development. Fig. 3 shows the theoretical dependence of the function $(1/m)\,(dQ/dt)$ on time; the specific heat generation of the animal organism increases from the moment of fertilization of the egg cell, has an S-shaped rise and reaches maximum value by approximately the end of embryogenesis, then it drops slowly during the growth of the animal and reaches a steady state in the adult organism. Precisely this course in specific heat generation has been observed experimentally in all species of animals [12; 6]. Hence, the result of our information-thermodynamic analysis of the process of ontogenesis is the solution of a problem that has thusfar remained unsolved: the unique course of the specific heat generation of the animal organism in time.

We have examined the historical sources of information as a factor with a physical value, and have shown that there exist three different approaches to the concept of information:

mathematical, physico-theoretical and biological. Biological information is the process of ordering that connects the micro-event – the signal with the macro-phenomenon – the objective. By applying the information-thermodynamic analysis we have established the laws governing the processes in the biological macromolecule hemoglobin, in the cell in a state of basal metabolism, and in the ontogenesis of the animal organism.

References

[1] G. Antonini, J. Wyman, M. Brunori, E. Bucci, C. Fronticelli, A. Fanelli Rossi on Symposium on the Structure and Activity of Enzymes. London 1964.

[2] G. Antonini: Interrelationship between structure and function in hemoglobin and myoglobin. Physiol. Reviews *45* (1965), 1, p. 123.

[3] L. von Bertalanffy: Theoretische Biologie, Bd. 1, 2. Bornträger, Berlin 1932, 1942.

[4] L. von Bertalanffy: General System Theory. George Braziller, New York 1968.

[5] L. Brillouin: Science and Information Theory. Academic Press, New York 1956.

[6] E. Calvet et H. Prat: Microcalorimétrie. Applications physicochimiques et biologiques. Masson, Paris 1956.

[7] B. D. Coleman, W. Noll: Conditions for Equilibrium at Negative Absolute Temperatures. Phys. Rev. *115* (1959), 2, p. 262.

[8] A. F. Cullis, H. Muishead, M. F. Perutz, M. G. Rossmann and A. C. North: A Three-dimensionals Fourier Synthesis of 5,5 Å resolution: description of the structure. Proc. Roy. Soc. (Ser. A) *263* (1962), p. 161.

[9] Th. de Donder, P. van Rysselberghe: Thermodynamic Theory of Affinity. Stanford University Press, Stanford 1936.

[10] E. Hecht: Thermodynamic Potentials for Systems at Negative Absolute Temperatures. Phys. Rev. *119* (1960), 5, p. 1443.

[11] R. Lemberg and J. W. Leppe: Hematin Compounds and Bile Pigments. Interscience Publ., New York 1949.

[12] J. Needham: Chemical Embryology. Vol. 2, Cambridge 1931.

[13] M. F. Perutz: Arden House Conference on Hemoglobin, Proteins and Nuclein Acids. Elsevier Publ., Amsterdam 1962.

[14] M. F. Perutz: Structure of Hemoglobin. In: Protein Structure and Fuction—Brockhaven Symposia in Biology. Upton, New York 1960.

[15] N. F. Ramsey: Thermodynamics and Statistical Mechanics at Negative Temperatures. Phys. Rev. *103* (1956), 1, p. 20.

[16] L. Szilard: Über die Entropieverminderung in einem thermodynamischen System bei Eingriffen intelligenter Wesen. Z. Phys. *53* (1929), p. 840.

[17] C. E. Shannon: A mathematical theory of communication. Bell Syst. Techn. Journal *27* (1948).

[18] K. S. Trincher: Biology and Information—Elements of Biological Thermodynamics. Consultants Bureau, New York 1965.

[19] K. S. Trincher: Thermodynamics of the elementary biological act (in Russian). Izvestiya akademii nauk, ser. biolog. (1968), 6, p. 907.

[20] K. S. Trincher: Zur Thermodynamik der Konformationsänderungen des Hämoglobinmoleküls. Zeitschrift für Naturforschung 17 b (1972), 5, p. 587.

[21] K. S. Trincher and A. G. Dudoladov: Spin-Lattice Interaction of Water and Protein Membranes in Cell Metabolism. J. theor. Biol. *34* (1972), p. 557.

[22] N. Wiener: Cybernetics. J. Wiley & Sons, New York 1948.

[23] J. Wyman: Linked functions and reciprocal effects in hemoglobin: a second look. Advances Protein Chem. *19* (1964), p. 223.

BIO-ECONOMIC ASPECTS OF ENTROPY

Nicholas Georgescu-Roegen

I. Introduction: Mechanistic Dogma and Economics

Man, in his continuous struggle to understand what nature is and how it works, has continuously sought support in some particular epistemological belief, some particular scientific dogma. A succession of such dogmas has punctuated the evolution of human thought with periods of epistemological fashions and it will continue to do so. During each of these periods, scientists have not only striven to accumulate evidence in favor of the dominant dogma but have also looked slavishly upon it as the only source of fertile inspiration. A glaring illustration of this dogma worship and especially of its possible consequences is offered by the economic discipline, which happened to come of age just as the mechanistic dogma was enjoying its apogee. That dogma had already exercised an exceptionally strong rule over scientific thought for more than three hundred years. But soon thereafter – about a hundred years ago – it was spurned by physics itself for reasons proper to that special science [10, pp. 69 – 122; 4]. Yet we still cling to it, albeit by the back door. There are good reasons for this stubborn attachment of the human mind to mechanics, in essence, to locomotion.

The idea that mechanics opens a vast avenue for man's arts has very old roots. The fascinating clocks that adorned numerous cathedrals and palaces heralded it daily long before Leonardo da Vinci argued in his *Macchine per volare* (at the opening of the sixteenth century) that man should be able to reproduce the mechanism represented by a flying bird. Descartes, who during the next century argued in his *De l'homme* that "the living body is a machine... not more, not less than the movement of a clock or of any other automaton," merely cast into explicit form a thought which had long since become an elementary article of faith for many. But without Copernicus, Kepler, Galileo, and Newton – to mention only the earliest architects of classical mechanics – the mechanistic dogma would not have acquired the supremacy in science and philosophy which Laplace expressed so poignantly in his famous apotheosis of mechanics [29, p. 4]. Everything in the world, whether in the past, the present, or the future – he proclaimed with a pride that reflected the temper of all scientific circles of that time – is completely determined by the basic laws of mechanics. Laplace did recognize that only a demiurgic mind could effectively determine the initial conditons of every particle in the universe and also solve the colossal system of equations which governs the motions of these particles. However, repeated situations in which the prediction of a particular event depends only on the solution of a few equations vindicated the mechanistic dogma in an increasingly spectacular way. The most spectacular of all such favorable testimonies occurred in 1846 when, less than one month after Urbain Leverrier announced before the French Academy both the existence of a new planet and its probable position, the observations of C. J. Galle from the Berlin Observatory confirmed *in toto* Leverrier's paper-and-pencil results.

Acting (as they all did) during the nineteenth century, the pioneers of economic science were led to look up – each in his own way and with varying degree of awareness – to mechanics as the model for any discipline deserving the name of science. Many admitted openly that economics cannot be conceived otherwise than as "*the mechanics of utility and self-interest*" – the clear-cut form in which Jevons [25, p. 21] defined that position. But the fact that the main body of economics has to this day remained completely faithful to the old mechanistic position does not have an equally simple justification. To be sure, there is the fact, on which Lord Kelvin insisted, that the human mind understands any phenomenon better if it is portrayed by a mechanical model. After all, human nature is such that we can act only by pushing or pulling. But this shortcoming of ours is no reason for science to remain always bound by this limitation. An economy of effort, badly misplaced, seems a more plausible cause of the epistemological inertia that characterizes modern economics. Indeed, any model involving, as all economic models do, a principle of conservation[1]) and a maximization rule is a mechanical analogue of the simplest possible structure, an analogue of a system reduced to its kinematic aspects. Actually, most economic models are not even kinematic, for they are not related to time in any precise way. With these models, the door was opened for an abuse of abstraction; as a result, economic theory has become the happy hunting ground for the lover of pure mathematical exercises. This is most unfortunate, for, as none other but a consummate engineer brought it into open [26, p. 111], mathematics may also serve as a substitute for the usually arduous task of coming to grips with the actual phenomena.

The blind tenacity with which modern economists have defended their worship of the number is unique in the annals of science. Happy within their own niche – an economic science reduced to paper-and-pencil exercises about ultra-simplified kinematic models – modern economists have lost any interest in following the evolution of other disciplines, including mechanics itself. Had any prominent economist given some thought to the subsequent fate of that science, economics in all probability would have gone through a crucial turning point long ago. For no sooner did Jevons and Walras make known their blueprints for the new sister of mechanics than the handwriting on the wall presaged the downfall of the mechanistic dogma.

[1]) Such as "Nothing is destroyed or created, everything is transformed at some specific rate".

II. Economics and Evolution

There are several salient consequences of the mechanistic nature of modern economics. There is first the standard analysis which reduces the market process to a pendulum analogue. No matter how prices or demand and supply schedules change, the initial situation is always restored *in toto* as soon as these changes are reversed. Everything follows the story of the famous Mother Goose rhyme in which the brave Duke of York marched his troops up the hill and marched them down again, without making any history. Alfred Marshall, who alone among the great insisted that economics is a sister science of biology rather than of mechanics, did point out that supply schedules undergo irreversible changes if demand schedules shift [30, p. 808]. Very simply, the same irreversibility was also shown to apply to demand schedules [12]. But modern economists, in general have preferred not to complicate their comfortable kinematic framework with such evolutionary phenomena.

An even more striking symptom of the kinematic epistemology is the conception of the economic process as a circular flow within a completely closed system. Witness the circular diagram connecting "production" and "consumption" by which this process is portrayed in practically every standard textbook.[2]) However, not even the circulation of money, taken in isolation, fits this picture, for even money (in whatever form) eventually wears out and must be replenished from sources extraneous to the money flow. It is, however, almost certain that the circulation of money has loomed large – as it still does – in the basic views of practically every economist [13, p. 8].

This circular concept of the economic process denies from the start that mankind's struggle with the material environment may have any role in the economic process. Had economists recognized the true nature of that process – which as I shall argue in this paper is entropic, not kinematic – the problem of pollution, which struck us suddenly in the face years ago, would have had another history; and so would the other side of the entropic problem, namely, the inconsiderate wasting of natural resources.

But if we go back to the dawn of the economic science, we do find there the recognition that labor is the father and nature the mother of wealth, as Sir William Petty, that keen student of human society, preached [24, II, p. 377]. All the more curious is it that, even though our knowledge of the past has increased and spread continuously ever since, modern students of economics have continued to remain indifferent to several glaring examples in which the environment has been the main factor of history. One plain and also well-known case is the Great Migration caused by the exhaustion of the fertility of the soil in Central Asia through long and ever increasing grazing. Remarkable civilizations – such as the Maya – crumbled away from history because of the inability of their people to migrate or to deal adequately with shrinking environmental resources. But above all there is the well-known (and indisputable)

[2]) To cite only a few of the most appreciated manuals: G. L. Bach, *Economics*, 2nd edn., Englewood Cliffs, N. J.: Prentice-Hall, 1957, p. 60; Paul A. Samuelson, *Economics*, 8th edn., New York: McGraw-Hill, 1970, p. 72; Robert L. Heilbroner, *The Economic Problem*, 3rd edn., Englewood Cliffs, N. J.: Prentice-Hall, 1972, p. 177.

fact that even in modern and contemporary history the struggle over the control of natural resources has played a dominant role.

Those who cling to the mechanistic dogma in the life sciences fail to see that history is evolutionary, i.e., it goes on in an irreversible, nay, irrevocable way, and only locomotion is completely reversible. Our planet, for example, could very well have traveled on the ecliptic in the opposite direction; no law of mechanics would have been violated thereby. The equations of mechanics are not disturbed at all if the time coordinate is changed from t to $-t$. Mechanics cannot, therefore, account for the unique direction in which evolutionary phenomena move. There is absolutely no way of ascertaining whether a movie of a bouncing, perfectly elastic ball is being run forward or backward, unless the scene also includes some irreversible processes. Nor is it possible for mechanics to account for qualitative change—which constitutes another fundamental aspect of evolution. Locomotion represents only a change of place, and such a change can always be reversed. On the other hand, even the simplest action of a human being, whether he burns a piece of coal or even whether he takes a breath, causes an irrevocable qualitative change in the entire system (consisting of that person and his environment). If modern economics is not an evolutionary science (as it should be), it is because its foremost servants have clung to the old Laplacean doctrine of a world totally determined by mechanics and only by mechanics.

III. The Entropy Law and Bio-economic Phenomena

Some economists have noticed that man can neither create nor destroy matter or energy [e.g., 30, p. 63], a truth which follows from the Principle of Conservation first established in mechanics and later adopted as the First Law of Thermodynamics. With this principle, therefore, we are still in mechanics and all is in order for modern economists to argue that man can produce only utilities by merely rearranging things. It is curious, however, that no one seems to have realized the obvious paradox in this position: why cannot man then rearrange the ashes and the smoke produced by a fire so as to reconstruct the original log?

To set the issue in plain light, let us consider the economic process in its entirety and view it only from the *purely material* viewpoint. The picture is that of a partial process which, like all such processes, is *analytically* circumscribed by an abstract, vacuous boundary which merely separates it from the rest of the material universe, i.e., from mankind's environment. What the economic process *does* consists then of the exchange of matter and energy across its boundary.[3]) This is what mechanics teaches us. But economics is not mechanics, not even pure physics. The fiercest defender of the position that environmental resources have nothing to do with value could not possibly deny that between what goes into the process and what comes out of it there is a *qualitative* difference. Obviously, the difference could not be quantitative: that would violate the Principle of Conservation.

[3]) For a detailed discussion of the analytical representation of a process, see [14, ch. IX].

An unorthodox economist would say that what goes in is *valuable forms of matter and energy*, valuable resources, and what the process discards can, generally, be only *valueless forms of matter and energy*, i.e., waste. And – a highly important point – this difference is by now recognized (although in different terms) by a particular (and peculiar) branch of physics, by thermodynamics. For thermodynamics, the matter-energy that goes into the economic process is in a state of *low entropy* and that which comes out of it is in a state of *high entropy*.

The concept of entropy, in all its details, is so involved that an authority on thermodynamics believes that it is "not easily understood even by physicists" [19, p. 37].[4]) To complicate matters, the term now circulates with many other meanings of a nonphysical, purely formal nature [14, App. B]. A recent edition of *Webster's Collegiate Dictionary* (1965) has three entries under entropy. Moreover, the definition pertinent to the physical world is utterly confusing: "a measure of the unavailable energy in a closed thermodynamic system so related to the state of the system that a change in the measure varies with change in the ratio of the increment of heat taken into the absolute temperature at which it is absorbed." But an older edition (1948) supplies a simple, intelligible definition which, although it cannot satisfy the specialist, would do for general purposes, such as those of the present topic: "A measure of the unavailable energy in a [closed] thermodynamic system." We need only add that energy exists in two qualitative states: *available* or *free* energy, over which man has appreciable control (more precisely, which can be transformed into work), and *unavailable* or *bound* energy, which man cannot use at all (which cannot be transformed in any way).[5]) The chemical energy of a piece of coal is free because man can use it for some purpose or other. But the vast amount of heat-energy dissipated in the sea waters is bound energy. Sailing ships cannot use it; instead, they need the free energy of a fuel or of the wind.

For free energy to exist, we must have a differential level, like that of a waterfall or that between the temperatures of the boiler and the cooler in a steam engine. As Nicholas Sadi Carnot first observed in a pathbreaking memoir on the efficiency of steam engines (1824), no engine can work if the same temperature, however high, prevails everywhere.[6]) Bound energy is the form in which energy consists of completely dissipated heat. Another way of describing the two concepts is to note that free energy implies order – the hotter steam in the boiler, the colder vapors in the cooler; whereas bound energy is characterized by disorder, by chaotic dissipation. Entropy, therefore, may also be viewed as an index of disorder. For this reason, we can say, for example, that the entropy of a copper sheet is *lower* than that of the ore from which it was produced.

The distinction between the two qualities of energy, which is specific to thermodynamics, is certainly an anthropomorphic one. But the point should not trouble any student of man,

[4]) The discussion of the Entropy Law in [21, p. 17] tends to vindicate this judgment. Even the notion of heat raises some highly intriguing issues, with the result that some physicists may go wrong on it, too. See *Journal of Economic Literature X* (Dec. 1972), p. 1268.

[5]) Technically, there is a difference between free and available energy and, *ipso facto*, between bound and unavailable energy. But this technical detail does not affect the present argument.

[6]) There is thus no sense whatever in the idea of [21, p. 17] that temperature measures "the usefulness" of heat.

nay, even a student of inert matter. We cannot get into mental contact with actuality except through anthropomorphic channels. If anthropomorphism is more striking in the case of thermodynamics, it is because the roots of this science lie in the common distinction between things that have economic value and those that are valueless waste. Indeed, thermodynamics grew out of Carnot's memoir, the first scientific work pertinent to the economy of scarce resources. Thermodynamics thus began as a physics of economic value and has remained so in spite of its subsequent developments of a more abstract nature.

A still stronger emphasis should be placed on the essentially economic nature of the Entropy Law, alias the Second Law of Thermodynamics. It is this law alone that distinguishes thermodynamics from mechanics. Indeed, it states that the universe or any other closed material system[7]) is at all times subject to an irreversible (rather, an irrevocable) process which is manifested by the steady increase of the entropy of the system. In other words, in any closed system there is a permanent tendency of free energy to turn by itself into bound energy, for ordered structures to turn by themselves into disordered ones. Briefly, any closed system continuously undergoes a process of irrevocable degradation the ultimate outcome of which is the complete dissipation of energy into uniform heat and of ordered matter into disorder. The ultimate state of the universe – so it appears – is the so-called Heat Death, but Chaos is a better term for it. Be this as it may,[8]) from incontrovertible and repeated experience we know that at least in our part of the universe and in the present cosmic phase the ice cubes in a glass of water will ultimately melt and the water in the glass will ultimately reach its thermodynamic equilibrium in relation to its surroundings.

Most important of all are two other points implied by the Entropy Law. First, the free energy of a piece of coal will ultimately be degraded into bound, useless energy even if, instead of using it, we leave the coal in the ground. Second, this degradation is irreversible or, rather, irrevocable. The upshot is that once the ice cubes in a glass of water have melted, under no condition would they form again by themselves. For if they would, entropy would decrease to a previous, lower level. Of course, we can cause all the water to freeze but only by degrading some free energy within a larger system that includes also the electrical plant which supplies the refrigerator with electricity. In this case, the Entropy Law requires that the entropy of this larger system increase. In straightforward arithmetics this means that in order to decrease the local entropy of the glass of water, the entropy of the rest of the system must be increased by a greater amount.[9])

[7]) A system is closed if no exchange of matter or energy occurs across its boundary. Obviously, the amounts of matter and energy in a closed system are constant. But the constancy of these amounts does not guarantee the increase of entropy; the amount of bound energy may decrease if there is exchange.

[8]) The issue is associated with the endless controversy surrounding the evolution of the universe [14, ch. VIII].

[9]) Here we can mention only in passing that the sustained efforts of the adepts of the mechanistic dogma ultimately brought mechanics back into thermodynamics in a new company: probability. In this process, however, "probability" has been mishandled as never before or since. The most important fact is that for the new discipline, for the so-called statistical mechanics, the spontaneous formation of ice cubes in a glass of water is not impossible; it is only highly improbable. This position has fostered the belief that, since the event is possible, by some ingenious device we may make it happen as often as we please, just as a sharper may throw a "six" any time he wishes. As P. W. Bridgman wittily put it, one could then "bootleg entropy" [14, ch. VI].

Economists are fond of saying that "there is no free lunch", everything must be paid its value so that price and value always balance out. The Entropy Law teaches us that mankind lives under a harsher commandment: *in entropy terms*, the cost of a lunch is greater than its price. It is regrettable, therefore, that a misconceived sales talk has led nuclear scientists to veil this truth. Even in the special literature, the nuclear reactor which, with the aid of some nuclear fuel, transforms nonfissionable but fertile into fissionable material now goes by the name of "breeder." What is more, the breeder is currently extolled for being able "to produce more fuel than it consumes," a description which is apt to mislead the unaware. Would it be accurate to say *only* that a hammer plant, for example, produces more hammers than it consumes? The stark truth is that even a chicken farm is not a "breeder;" according to the Entropy Law, even such a process consumes more free energy than is contained in its product. The confusion that, in this or some other manner, has been allowed to spread among laymen constitutes a great obstacle to a correct understanding of the environmental problem. Certainly, the call "to improve the environment," which a prestigious organization such as the United Nations repeatedly made in its Stockholm Declaration on the Human Environment (June 1972), is likely to foster the impression that it is in our power not only to stop the working of the Entropy Law, but also reverse it.

Some—Henri Bergson in particular [3, pp. 245f]—have claimed that life is an exception to the above commandment, that it evades the entropic degradation. Do not all organisms seem to maintain their highly ordered structure intact at least over some finite periods of time? Naturally, the claim cannot possibly mean that life can defeat the Entropy Law. As one great natural scientist after another has pointed out, life, although not fully explainable by the laws of matter-energy, cannot possibly go counter to any of them. What an organism does on the whole is simple; as Erwin Schrödinger put it in a great little book [33, pp. 71f], it sucks low entropy from the environment (and eliminates high entropy). In this way, the organism strives to compensate for the steady, inevitable decay of its delicate structure. The obvious similarity of this process with the economic process viewed as a whole vindicates Marshall's intuitive perception of the kinship between the two. Clearly, maintenance is the undeniable purpose of both processes.[10])

The presence of life does, however, matter. As the simplest facts show, life influences the entropic degradation, at times, by retarding it—as when green plants prevent part of the solar radiation from being degraded *immediately* into dissipated heat—at others, by speeding it up—as when animals feed on vegetation or on other animals. There is no crass vitalism in this position. For—a point worthy of unparsimonious emphasis—the Entropy Law imposes neither a definite speed nor a particular pattern on the entropic degradation. All it says is that such a degradation is unavoidable and irrevocable. It is because of this entropic freedom of actual phenomena that life can don an unlimited variety of forms and we can speed up the entropic degradation of the environment according to the economic plans of our own choosing [14, pp. 12, 194f].

[10]) There are several other considerations that lend full support to the view that the economic process is the extension of the biological [14, pp. 307f, 348f].

It would be absurd, however, to see in a life process only the transformation of low into high entropy, of environmental resources into waste. Waste is only a material by-product of such a process. Its true output is the enjoyment of life, of which we are fully conscious and which must exist as an entropic flux at every biological level. On the other hand, it would be equally absurd to ignore completely—as modern economists do—the continuous and irrevocable tapping of environmental resources as well as the uninterrupted flow of polluting waste. These are not only irreducible aspects of the economic process but also the factors that mold its evolution. It is because of the scarcity of the resources needed for the maintenance of the human species that man, ever since he discovered his industrial abilities, has striven to discover new forms of free energy as well as more efficient means of utilizing them.

IV. The Entropy Law and Economic Scarcity

The taproot of economic scarcity lies in the Entropy Law. It is because of this law that, jus as no living organism is immortal, no ordered structure can last forever. True, any ordered structure can be recreated, but only by an additional degrading of free energy such that the final balance must show an entropy increase. And since free energy itself can be used only once, the same may be said of any ordered structure.

Imagine that, on the contrary, the copper roof of a building would last indefinitely and the free energy of a ton of coal could be used over and over again. Then mankind would not have to keep mining copper and coal continuously, but only to the extent to which *new* needs for these resources would emerge. Scarcity of environmental resources could then arise only if their amounts were limited, in which case scarcity would be identical to the actual scarcity of Ricardian land (i.e., of land as pure space). Ricardian land, though limited, can be used over and over indefinitely. In this imaginary world, the size of the population at any one time would be limited by the total *fund* of resources, but the life span of the human species would know no limit. In the actual world, however, the Entropy Law sets a finite limit to this life span, too (Sec. VI, below). Physicists, such as H. von Helmholtz, and biologists, such as J. B. S. Haldane, have repeatedly warned their human fellows that the most certain fate of mankind is that of any other species: extinction.

This conclusion certainly offends man's self-attributed superior dignity. Man has always had an inner compulsion to think of himself as a unique, extraordinary creature, above the laws applicable to other species. In Genesis man proclaims himself to be the image of God Himself. He once believed that the entire universe turns around his petty abode, later, that only the sun does so. At one time, man entertained the myth of perpetual motion of the first kind, which would make him almost independent of any exterior power because he could then use the same amount of energy over and over on end. Later he hung on to the myth of perpetual motion of the second kind, in which free energy can be switched between heat and work, back and forth, without any degradation. Both myths still survive under the widespread, albeit dormant, attachment to the feasibility of "entropy bootlegging." A kindred myth has been brought to light by the recent controversies surrounding the menace of pollution and the energy crisis. According to it, man will forever be successful in discovering new and ample sources of energy and in inventing increasingly efficient means of harnessing them. Thus, although man as an individual is not immortal, at least his species is so.

We can no longer doubt that it is because economists, in general, shared this myth that the standard picture of the economic process completely ignores the environmental factor. For, now that the sterility of that picture has come to be exposed by events as well as by critics, standard economists are set on defending that myth explicitly, and at any price. They, who have always decried any institutional or historical approach in economics, do not scruple now to invoke man's past technological successes in defense of the standard framework [2; 28; 39]. W. Beckerman, for example, hammers on the fact that economic growth has never ceased since ancient times [2, pp. 332f].[11]) The rub is that in the case of a unique specimen—as mankind is for us—the historical argument is not convincing. Cancer does strike one in spite of one's perfectly healthy past. In fact, the historical argument may cut both ways. And history also shows that life-supporting sources have dried up and some large communities have succumbed to the shrinking supply. In situations of this sort, other considerations may be more reliable for showing on which side the truth rests.

Our spaceship—the earth—floats, as it were, within the cosmic store of free energy. Unfortunately, only an infinitesimal part of this energy is accessible to man. For man will remain confined to a speck of the cosmic space even if he could travel in space with the speed of light; just to scout the neighborhood of the nearest sun for some possible, yet uncertain, earth-like satellites would take nine years. Intersidereal travel has no economic promise [8, p. 28]. Man's biological nature sets even narrower limits to the energy which he can handle. Too high or too low a temperature is incompatible with his survival. And so are many radiations. Briefly, both the space and the amount of free energy at our disposal are finite. *Ergo*, the human species must come to an end. No amount of confidence in human nature and no assurance from a professional chair that economic growth will continue indefinitely [2, p. 344], can change this fate.

Just as in the case of an individual organism, the natural end will be brought about by a very slow aging process, which occasionally may manifest itself by some visible crisis. But especially a species such as ours will certainly not succumb to the first crisis. Nor is any crisis likely to span only a few years, but centuries or, rather, millenia. The factors responsible for the aging process work continuously, even when no palpable symptoms betray their presence. Moreover, their work is in one direction and irrevocable. A living creature—as a Jain philo-

[11]) History is invoked also by many natural scientists who appear to share the myth of an immortal mankind [1]. But the economists' reaction has other tale-telling aspects. For years now, standard economists have extolled the econometric model as the only reliable tool of scientific prediction. They have also used models with a constant growth rate almost exclusively and, moreover, have preached that an economic plan is successful only if it can maintain an exponential growth. Yet standard economists have assailed the recent report of the Club of Rome [31] for using these very tools of their trade. And even that is not all. Highly respectable writers and publications have not hesitated to resort to a pollution of insults, plain or veiled, against those who—like "the whiz-kids from MIT" [2, p. 327]—have tried to bring up the ecological problem.

sopher observed – begins to die at birth. The recent pollution calamities and the looming energy crisis very probably are the first clear foretokens of old age. But this does not mean that the end is near, nor even in sight. Actually, we have no means of predicting even its probable timing. That is why one may fault the report of the Club of Rome for predicting that mankind will reach a catastrophic crisis, a point of no return, within the next four decades [31]. But the report nevertheless has the merit of exploring systematically some basic aspects of mankind's slow march to extinction. And there is no sense in trying to set it aside by looking only a few decades ahead so as to be able to conclude that mineral resources suffice to sustain the prevailing level of affluence in the USA [e.g., 18].

V. Solar Versus Terrestrial Energy

The simple observation that our environment is finite only grazes the economic problem of mankind. All species live in a finite environment. Also, practically every species – including man – derives its biological existence, directly or indirectly, from the solar radiation that reaches our planet. Man is unique in that, in addition to the endosomatic organs (organs with which an organism is endowed biologically), he uses exosomatic organs (artifacts that are usually produced socially). The exosomatic organs allow man to use other forms of free energy than solar radiation, namely, the free energy stored in the bowels of the earth. Their production, in turn, calls for the use of various mineral resources. Because of the extension of the endosomatic evolution into the exosomatic evolution, man can fly although his body has no wings and can swim under water although his body has no fins and no gills [14, pp. 307f]. Most important of all, he can do all these things by spending only a very small amount of his biological free energy, that is, without experiencing the great discomfort that usually accompanies physical exertion. The fact is that the human species has thus become addicted to industrial comfort and, from all we know about evolution, such an addiction is irrevocable. Indeed, it is difficult to imagine that mankind may revert to a berry-picking economy.

With the exosomatic evolution, the existence of the human species came to depend on two distinct kinds of free energy, each coming from a different source – the terrestrial and the solar. And the crucial point for the economics of resources is that these two sources present a series of striking asymmetries.

First, the terrestrial energy is a *stock*; that of the sun reaches us as a *flow*. The difference should be well marked. Coal-in-the-ground is a stock because we may use it all within a year (conceivably) or over centuries, if we wish. By contrast, we cannot use today any part of tomorrow's solar radiation any more than we can use today the future services of, say, a building [14, pp. 226f]. While we are to a large extent master over the rhythm in which we may use terrestrial low entropy, the flow rate of solar radiation is beyond our control. This flow is strictly determined by the basic coordinates of our solar system, above all by the size of the globe. Ricardian land has economic value because it is the net which catches the most valuable thing – solar radiation – and, in addition, its size is immutably fixed.[12])

Second, it is solar energy, over which man has no control, that is indispensable for all life on earth. On the other hand, man could conceivably live without any tapping of the energy he can control. True, man can increase his benefit from solar energy – as by forcing land to yield larger crops, by using the power of waterfalls, or by using solar batteries – as a rule, only with the aid of some input from his terrestrial dowry.

Third, man continuously competes, at times and places even fiercely, with other species for solar energy (as such, or in some ulterior form). In this competition man, too, has followed the law of the jungle, made even more merciless by his highly sophisticated exosomatic weapons. Man has sought openly to exterminate rabbits, wolves, weeds, insects, microbes, etc. – species that either steal his food or feed on him. His struggle for food has brought some species (the whale is a good example) to the brink of extinction, while others are menaced by his craving for the extravagant. Still more significant, because it has been done unobtrusively, is the slow extermination of some draft animals, such as the water buffalo and the Percheron horse. Not being able any longer to produce enough food *and* fodder on the increasingly crowded land, peasants have done away with the water buffalo and turned to using a tractor instead.

By contrast, man has no significant competitor for mineral resources. Yet even industria activity may endanger other species as well as man. To be sure, man must see to it that unnecessary harm is avoided and the species beneficial to him are protected. But some environmentalists stretch the argument beyond reason by clamoring, for instance, that the Alaskan moose must be preserved even at the cost of unheated hospitals and homes. There simply is no law of evolution saying that no species should be forced out of existence by others. The contrary is true. If pressed for survival, no living creature will simply lie down to die so as to let others live; all creatures will strive for survival without concern for that of others.

Fourth, as the Entropy Law teaches us, pollution is an indispensable by-product of any industrial activity using terrestrial low entropy. All pollution is, in various degrees, harmful to life. It consists of some irrevocable chemical or nuclear alterations of the environment and, in all cases, of some increase of the environmental temperature.[13]) "Bigger and better" refrigerators, automobiles, jet planes, etc. necessarily mean not only "bigger and better" depletion of exhaustible resources but also "bigger and better" pollution. Mankind is like a household which uses things from a limited pantry and throws the inevitable waste into a finite trash can. The fact that we have come to realize the problem of pollution but are still reluctant to see the correlative phenomenon (the exhaustion of terrestrial resources) is understandable. Pol-

[12]) Those who, like Carl Kaysen [28, p. 663], insist that "new land can be created by new investment" simply run away from the above truth.

[13]) Thermal pollution is a most crucial one. The heat discharged by a nuclear plant may raise the temperature of the Hudson River by some 4° F with the best arrangement, otherwise by as much as 7° F. Where to build the next plant, and the next, is a formidable problem. The environmental temperature may increase also because, as more carbon dioxide is accumulated in the atmosphere, the greenhouse effect of the atmosphere is intensified. Should the ice caps melt, large continental areas will be submerged. To trust the specialists [34, pp. 56—58; 7, p. 152; 40], the problem is far from being "an old scare", as Beckerman dismisses it [2, p. 340].

lution is a surface phenomenon; moreover, most of it is felt by the generation which produces it.[14]) Depletion involves volume and, as a rule, affects the generations of some distant future.

In contrast with the terrestrial low entropy, solar energy is *pollution-free*. The solar energy which reaches the earth and is not used by photosynthesis or transformed into the movement of the air or into the evaporation of water ultimately becomes dissipated heat. But this heat is extremely useful; it keeps the thermal equilibrium between the globe (at the proper temperature) and outer space. The situation is not changed if any solar energy is transformed into work. And since dissipated heat is the only waste of such a transformation, the harnessing of solar radiation will produce no pollution at all.[15]) But there are still stronger reasons why solar energy must occupy a vital role in the long-range economy of mankind.

Fifth, the difference between the size of the terrestrial stock and the solar flow of free energy is of astronomical proportions. The sun radiates 10^{14} Q per year, at the cost of a loss of mass of 131×10^{12} tons. Of this fantastic amount (one single Q is equal to 10^{18} BTU) only some 5,300 Q reach the globe. But at our scale even this last amount is stupendous. Photosynthesis over the whole globe uses only 1.2 Q; even the total energy consumed annually by mankind currently amounts to not more than 0.2 Q. Moreover, the sun will continue to shine with practically the same intensity for another five billion years (before becoming a red giant that will completely scorch the earth at 1,000° F). Undoubtedly, mankind will not live to need all this abundance. The opposite is true for terrestrial resources. If mankind's dowry of fossil fuel, which initially amounted to 215 Q, had been burned all at once, it would have produced only two weeks of sunlight![16]) Such meager reserves cannot last more than a few decades if demand continues to increase at the current rate.

Hopes have turned in many directions, first of all, to nuclear fuel. But even the reserves of uranium 235 would not last much longer if used in the ordinary reactor. Many hopes are now set on the breeder reactor which can also use the fertile but nonfissionable elements (uranium 238 and thorium 232). Some authors even claim that this kind of energy is "essentially inexhaustible" [36, p. 412]. Yet no estimate of the reserves, even rough estimates, are available. We are only instructed that large areas of the United States, for example, are covered with black shale or with granite containing about 60 grams of natural uranium per metric ton [22, pp. 226f]. The strip mining and crushing of these rocks would provide ample fuel for breeder reactors, which may yield from 2,000 to 3,000 times the amount contained in the initial fossil fuel resources. A grand plan by Weinberg and Hammond [36; 37], envisages the building of some 4,000 nuclear "parks" offshore, each containing eight breeder reactors. But, as responsible forums now admit, the breeder still presents risks of radioactive catastrophes; moreover, the problems of the safe storage of the radioactive garbage and of the cumulative effect of radiation are far from solved for any expanded use of nuclear energy [15; 16]. The

[14]) For an account of the various types of pollution documented with striking photographs, see Young [38].

[15]) This does not exclude the possibility of climatic changes due to large-scale alterations of the heat distribution on the globe, the so-called heat-island effect.

[16]) These computations are based on the data given by Hubbert [22]. The reader should note that some data vary greatly from one author to another, but not so much as to affect the general picture presented here.

plan raises numberless other problems—for example, that of the climatic alterations caused by so many heat islands (which is recognized even by the proponents).

There remains the physicist's dream, the controlled thermonuclear reaction. The deuterium-tritium reaction uses lithium 6. Since this element exists in a rather small quantity, the reaction could yield only about 200 Q, i.e., about the same amount of energy as that of the initial stock of fossil fuel.[17] It cannot therefore give rise to any sanguine hopes, although it stands a better chance to succeed because it requires a lower temperature than the fusion of deuterium alone. This second reaction is a different story. If only one percent of the deuterium in the oceans were to be used, it would yield 10^8 Q, an amount sufficient to support an energy consumption rate five times greater than the present one for a hundred million years. Unfortunately, in spite of sustained efforts to control fusion, success is not anticipated even by experts. The reaction, akin to the very source of the sun's energy, can take place only at temperatures higher than 0.2 billion °F, hotter than the interior of the sun. This is why fusion may never be harnessed within a material container.[18] We do have examples of energies—such as that of dynamite—which can be used only as a bomb.

To be complete, we should mention also the flows of tidal and geothermal energy. But, even though their total amount is not negligible (0.1 Q per year), there are very limited situations in which these flows can be exploited at all.[19]

To sum up, the energies of terrestrial origin on which we can effectively rely exist in very limited amounts; those which exist in ampler amounts are surrounded either by great risks or by hopeless technical obstacles. In contrast with this picture, there is the immensity, at our scale, of the energy received from the sun. True, the direct use of solar energy is not yet practical on a significant scale. But its feasibility is a proven fact. Solar energy is already being used in many situations; only the economic efficiency of its large-scale operation is still a desideratum. The main obstacle is that solar energy reaching the earth is very diffused and also very uneven. It is as if we had to operate a turbine, not by a waterfall, but by capturing the kinetic energy of an equivalent amount of rain drops.[20] However, news of promising results is coming from various directions.[21]

17) It certainly could not supply the world's energy for 120 million years, as Holdren asserts [21, p. 108].

18) For a brief review of the technical difficulties at the present moment, see [32]. We should also note that while fusion is almost completely free of radioactivity, the dissipated heat poses even greater problems than for fission.

19) For a convenient account of all energy sources, see [23; 24; 35].

20) Of course, the kinetic energy of rain drops, of waterfalls, and of the wind, derives from solar energy.

21) At the ground level, A. B. Meincl and his wife, Marjorie, have succeeded in converting solar energy into electricity with 30 percent efficiency [20; 23, pp. 66f; 35, pp. 677—681]. P. Glaser places new hopes on the use of artificial satellites for capturing the more intense solar radiation in outer space [17; 35, p. 681]. For various developments in the USSR, see Borissov [5].

VI. The Economics of Natural Resources

There are several reasons why the usual economic considerations cannot apply to natural resources. One reason is related to the economist's creed that "resources are properly measured in economic, not physical, terms" [28, p. 663], a creed adopted by almost every other writer on resources. But prices, far from having the power attributed by economists of even inducing the creation of land, energy, and materials, are parochial economic factors. The only relevant way to look at the supply of free energy in the long run is in real terms: any technique is viable provided it can produce an amount of *harnessed* energy greater than it consumes. Clearly, there it no sense in counting on the oil in some shale if no method can extract from that source more oil than it consumes. On the other hand, a technique for trapping solar energy, now economically inefficient, may become highly efficient at some future price constellation, provided that it is efficient (as most such techniques are) in real, as opposed to monetary, terms. The upshot is that solar energy cannot be ignored by the economics of resources. Solar energy is the most important factor in the long-run picture. For all practical purposes, solar energy is a free good; moreover, it does not even have a cost in environmental deterioration.

In the long run, it is the terrestrial source of energy which is the scarcer; but in the short run, it is the solar source. The fact that both the flow rate of radiation and the size of the net that catches it are immutably fixed sets a limit to "the amount of life," including the number of people that may exist at any one time on the globe. Naturally, this limit is influenced by many other factors, which is particularly true for the size of the human population.

The problem of the size of the world population is despairingly complex. Yet oversimplification characterizes most discussions of it. Some writers have claimed that the globe can feed as many as 45 billion people if the best farming methods are applied to every acre [6, p. 35]. Weinberg and Hammond claim that their own plan may support 20 billion people at the standard of living now prevailing in the USA [36, pp. 415f; 37]. More recently, an authoritative source has claimed that the same feat may be achieved even for a population of 3,000 billion [28, p. 664]. All these projections ignore the crucial fact that the problem of size is not reducible to that of a simple change of scale [14, pp. 105–107]. A ten times increase in the supply of food does not guarantee at all that a population ten times as numerous may exist. Beyond a certain limit, the organization of society encounters increasing difficulties of a nature akin to entropy. It appears highly doubtful whether an ordered organization of the complexity and efficiency required by the presumed multiplicative level of production would be achievable. Such difficulties will beset even a population double the present size, which is the tolerable maximum limit according to more sober analyses [8, p. 13; 11, p. 100]. For the technical progress and the increased population density have reversed the truth of the old Arab legend. Nowadays, it is easier to move the mountain to Mohammed than to persuade Mohammed to journey to the mountain. And the crucial rub is that some Mohammed must go to the mountain to move it.

We should also note that none of these projections has considered the question far more vital for mankind's future: How long can a given world population – be it of one million or of one hundred billion – be maintained?[22]) In the light of this question, several important issues become manifest.

Among the multiple aspects of man's competition with other creatures for the finite flow of solar energy, there is the fact that the number of people who can be fed at any one time depends (among a host of other factors) on how many draft animals are used in farming. Draft animals must also be fed; hence, food crops can be raised only on part of the available farm land. Organic agriculture, which uses only the power of draft animals and revives the soil only with manure, yields *ceteris paribus* the smallest food crop for any given area. To obtain the maximum food crop, one must turn to mechanized agriculture and to the new highyield varieties. The difference between the two types is crucial. The draft animals being solar energy converters, organic agriculture necessitates only a small input of terrestrial origin (for the various implements). By contrast, mechanized agriculture and high-yield varieties require a very great input of terrestrial energy and materials; the benefit from solar energy is confined only to photosynthesis (which, to be sure, is increased).

The conclusion, surprising though it may seem, is that the mechanization of agriculture and the introduction of the high-yield varieties constitute a move against the grain as far as the long-run economy of resources is concerned. It substitutes the scarce resources for that which is practically a free good.[23]) The move was, however, inevitable because of the ever increasing pressure of population on the annual flow of food. In spite of what one may hear rather often, the size of the population does play a unique role in the environmental struggle of man. Only some consequences can be explored here in a quick step.

Even if the direct use of solar energy could render mankind completely independent of any terrestrial energy, other elements of man's terrestrial dowry would still retain a limitative influence on the amount of human life. Because of our exosomatic evolution, we depend on the continuous supply of various elements – iron, copper, calcium, phosphorous, etc. But the corresponding resources are finite. On the other hand, these resources are subject to the work of the Entropy Law; they inevitably dissipate and the dissipation is accentuated substantially by their use by man. True, one may slow down the rate of dissipation by recycling. But recycling is neither free, nor perfect. Recycling consumes a substantial amount of energy and other resources, which must become infinite as the operation is made more intensive. In view of the Entropy Law, it is hard to think that recycling may stop dissipation entirely. One can pick up all the pearls of a broken necklace, but no actual process could possibly reassemble all the atoms of a used-up coin. With the best recycling, therefore, the stock of any element in use would still be depleted by some constant factor.[24]) Unfortunately, some elements are so dissipative that even recycling seems out of the question. The worst situation is that of the elements which, in addition, exist in a very limited supply. At least one vital element –

[22]) The Weinberg and Hammond plan may constitute an exception, in that the projected twenty billion population can presumably be maintained as long as the corresponding supply of nuclear fuel lasts—millions of times longer than the supply of fossil fuel [37].

[23]) The point explodes the ineptitude of the thought that the production of food protein from mineral oil on a large scale will completely solve the food problem.

[24]) Some data concerning recycling are given in Cloud [8, p. 14].

phosphorous – belongs to this category [18]. Energy may help us to reach deeper for minerals and to prolong the service of the crustal resources, but so far we have no basis for claiming – as Weinberg and Hammond do [36, pp. 412f; 37] – that energy is convertible into most life requirements. In the economics of resources, therefore, matter must be considered side by side with energy.

Let us simply denote the aggregate stock of *mineral* resources by S. If s is the average *net* amount dissipated for each living person per year, then S/s is the amount of people-years, viz., the amount of human life, that can be supported by S. If P is the average population, then an amount $r = Ps$ is consumed each year, and $t = S/r$ represents the "calculated" span of the exosomatic life of mankind in years [13, pp. 12f; 14, pp. 304].[25]) The greater is the annual decumulation r, the shorter is this span. Now, r may increase for two reasons. First, the average population may increase, which means that future lives are discounted earlier. Second, s may increase because the same population uses resources for satisfying man-made wants, which usually are extravagant wants. Sketchy though this picture is, two conclusions are indisputable.

First, broadly speaking, whether a baby is born now or ten thousand years from now, it will deplete the finite stock S by roughly the same amount. Consequently, the amount of life supported by S is not affected if P varies within the limit imposed by the supply of food. Two facts account for the widespread error that, on the contrary, the scarcity of natural resources arises from a continuous population growth. Growth has indeed been responsible for some recent incontrovertible symptoms of that scarcity. There is also the fact that the authors of *Limits to Growth* have singled out the constant growth rate as the culprit for the predicted ecological catastrophe and, as a result, have seen the salvation in a zero-growth population [31, pp. 163–175]. Others, too, have seen the salvation in John Stuart Mill's old notion of stationary state, not necessarily for the same reasons [8, p. 9; 9; 36]. Yet even zero-growth population cannot make more ample what is definitely scarce. The point that in a stationary state technology has more time to adjust itself to the resource depletion [31, p. 166] is valid only if the pressure of population, which sets the pace of depletion, is low. The same is ture about the point that some pollution has more time to be eliminated by natural factors. Possibly, one may also argue that in a growing system each generation has to educate a greater generation than itself and to accumulate capital for it; this would ignore that each generation works with an inherited increased capacity (which is the very advantage of a growing system).[26]) The basic truth is that zero-growth population cannot possibly make more abundant what is limitedly scarce. Actually, some of the arguments in favor of the stationary state lead straight to the conclusion that a decreasing population is the most desirable. Be

[25]) A similar coefficient is used by Cloud [8, p. 9], who calls it "the demographic coefficient."

[26]) The last argument may be related to Daly's position [9] that in a stationary world material growth would be replaced by moral growth—a highly desirable yet unproved prediction. One must be cautioned, in particular, against the more serious pitfall of invoking the Prigogine thermodynamic principle in support of the stationary state. The principle states that, under some special assumptions, the entropy produced by an open thermodynamic system *attains* its minimum when the system becomes stationary. However, the said assumptions are not properly met even by biological organisms [27, pp. 227—235]. The application to mankind's economy is sheer fantasy.

this as it may, a stationary state is incompatible with the actual world, where continuous depletion of energy and materials compels mankind to keep adjusting itself to evolving situations, sometimes for the better, sometimes for the worse.

The second conclusion is that every Cadillac, every Rolls-Royce, and every Zim, let alone every war weapon, mean fewer plowshares for some distant generations. Economic development through a fantastic industrial abundance may be a boon for us now and for those who will enjoy it in the near future, but it is against the interest of the human species as a whole.

Ordinary economics has frequently been defined as the study of the administration of scarce resources. We now see that one should have added "among the active members of the same generation." The economics of resources, on the contrary, raises the problem of their distribution among all generations. For this problem, the market or the plan of ordinary economics can be of no help to us. Neither the demand nor the supply of future generations is present on the market or in an economic plan. The farthest outlook in the recent ecological controversies has generally stopped at the end of this century. To find a means by which to bring the demand of future generations into the present defies the imagination, however well prepared. The problem boils down to convincing every individual to reduce his present high standard of living to such a low level that the generations of one million years from now should also enjoy the same comfort. When and if solar energy will become the only source of power, the economic dictatorship of the present over the future generations will lessen, but will not disappear entirely. After all, the destiny of the human species may be to have a short but thoroughly exciting and extravagant life. Let the less ambitious species have an uneventful but long life so that they may inherit the earth and continue to enjoy the power of the shining sun.

References

[1] Philip H. Abelson: Limits to Growth, Science (17 March 1972), p. 1197.

[2] Wilfred Beckerman: Economists, Scientists, and Environmental Catastrophe. Oxford Economic Papers *XXVI* (1972), 327—344.

[3] Henri Bergson: Creative Evolution. H. Holt and Company, New York 1911.

[4] R. J. Blin-Stoyle: The End of Mechanistic Philosophy and the Rise of Field Physics. In Turning Points in Physics (R. J. Blin-Stoyle et al., Eds.). North-Holland, Amsterdam 1959, 5—29.

[5] Oleg Borissov: Sun-raised Energy Warms Up. Energy International *IX* (June 1972), 25—27.

[6] Colin Clark: Agricultural Productivity in Relation to Population. In: Man and His Future (G. Wolstenholme, Ed.). Little, Brown, Boston 1963, 23—35.

[7] Preston Cloud: Mineral Resources from the Sea. In: Resources and Man. W. H. Freeman, San Francisco 1969, 135—155.

[8] Preston Cloud: Resources, Population, and Quality of Life. In: Is There an Optimum Level of Population? (S. F. Singer, Ed.). McGraw-Hill, New York 1971, 8—31.

[9] Herman E. Daly: The Stationary-State Economy. Distinguished Lecture Series No. 2, Department of Economics, University of Alabama 1971.

[10] Albert Einstein and Leopold Infeld: The Evolution of Physics. Simon and Schuster, New York 1938.

[11] David M. Gates: The Flow Energy in the Biosphere. Scientific American *CCXXIV* (September 1971), 89—100.
[12] Nicholas Georgescu-Roegen: The Theory of Choice and the Constancy of Economic Laws. Quarterly Journal of Economics *LXIV* (1950), 125—138. Reprinted in Nicholas Georgescu-Roegen: Analytical Economics: Issues and Problems. Harvard University Press, Cambridge, Mass. 1966, 171—183.
[13] Nicholas Georgescu-Roegen: The Entropy Law and the Economic Problem. Distinguished Lecture Series No. 1 (Delivered December 3, 1970), Department of Economics, University of Alabama. Reprinted in The Ecologist *II* (July 1972), 13—18.
[14] Nicholas Georgescu-Roegen: The Entropy Law and the Economic Process. Harvard University Press, Cambridge, Mass. 1971.
[15] Robert Gillette: Nuclear Safety: Damaged Fuel Ignites a New Debate in AEC. Science (28 July 1972), 330—331.
[16] Robert Gillette: Reactor Safety: AEC Concedes Some Points to Its Critics. Science *CLXXVIII* (3 November 1972), 482—484.
[17] Peter E. Glaser: Power from the Sun: Its Future. Science (22 November 1968), 857—861.
[18] H. E. Goeller: The Ultimate Mineral Resource Situation. Proc. Nat. Acad. Sci USA *LXIX* (October 1972), 2991—2992.
[19] D. ter Haar: The Quantum Nature of Matter and Radiation. In Turning Points in Physics (R. J. Blin-Stoyle et al., Eds.). North-Holland, Amsterdam 1959, 30—44.
[20] Allen L. Hammond: Solar Energy: A Feasible Source of Power? Science (14 May 1971), p. 660.
[21] John Holdren and Philip Herera: Energy. Sierra Club, San Francisco 1971.
[22] M. King Hubbert: Energy Resources. In: Resources and Man. W. H. Freeman, San Francisco 1969, 157—242.
[23] M. King Hubbert: The Energy Sources of the Earth. Scientific American *CCXXV* (September 1971) 61—70.
[24] C. H. Hull, Ed.: The Economic Writings of Sir William Petty. 2 vols. The University Press, Cambridge, Eng. 1899.
[25] W. Stanley Jevons: The Theory of Political Economy. (1871), 4th edn. Macmillan, London 1921.
[26] Reginald O. Kapp: Towards a Unified Cosmology. Basic Books, New York 1960.
[27] A. Katchalsky and Peter F. Curran: Nonequilibrium Thermodynamicsin Biophysics. Harvard University Press, Cambridge, Mass. 1965.
[28] Carl Kaysen: The Computer that Printed Out W*O*L*F*. Foreign Affairs *L* (1972), 660—668.
[29] P. S. Laplace: A Philosophical Essay on Probability (1814). John Wiley and Sons, New York 1902.
[30] Alfred Marshall: Principles of Economics. 8th edn. Macmillan, New York 1924.
[31] Donella H. Meadows et al.: The Limits to Growth. Universe Books, New York 1972.
[32] William D. Metz: Fusion: Princeton Tokamak Proves a Principle. Science (22 December 1972), p. 1274B.
[33] Erwin Schrödinger: What is Life? The University Press, Cambridge, Eng. 1945.
[34] Edward Teller: Energy Pattern of the Future. In: Energy and Man: A Symposium. (Courtney C. Brown, Ed.). Appleton-Century-Crofts, New York 1960, 55—72.
[35] Kenneth F. Weaver: The Search for Tomorrow's Power. National Geographic *CXLII* (November 1972), 650—681.
[36] Alvin M. Weinberg and R. Philip Hammond: Limits to the Use of Energy. American Scientist *LVIII* (1970), 412—418.
[37] Alvin M. Weinberg and R. Philip Hammond: Global Effects of Increased Use of Energy. Paper presented at the Fourth International Conference on the Peaceful Uses of Atomic Energy, September 7, 1971.
[38] Gordon Young: Pollution, Threat to Man's Only Home. National Geographic *CXXXVIII* (December 1970), 738—781.
[39] Limits to Misconception. The Economist *CCXLII* (11 March 1972), 20—22.
[40] The Fragile Climate of Spaceship Earth. Intellectual Digest *II* (March 1972), 78—80.

INFORMATIONAL PATTERN OF THE HIERARCHICAL SYSTEM OF CONTROL IN AN INTEGRAL ORGANISM

S. N. Braines and V. B. Svetchinsky

The problem of control in an organism is among the key issues in modern biology and medicine. Classical biology by and large concentrated on the laws behind the processes connected with the transformation of matter and energy in an organism. Material (i.e. pertaining to matter) and energy flows constitute the essence of any organism. Hence the natural and obvious urge to gain an insight into the mechanism of material and energy exchange. Far less obvious, though, is the fact that the material and energy exchange, which provides for the life activity of an organism, cannot proceed normally unless a certain organizing influence is exerted upon it, that is to say unless it is governed by what is known as a control process. The explanation of this phenomenon is simple enough: obeying the 2nd law of thermodynamics, material and energy processes tend to reach a certain state of equilibrium corresponding to the maximum entropy of the system (the word "system" herein stands to denote a closed system: "organism-environment"), whereas the life activity of an organism is a non-equilibrium process in a sense tending to offset the increase in entropy. The reason for such a peculiar behaviour of the material and energy processes of life activity should be sought in the pattern of their organization, in the control processes occurring in the organism. The latter, in their turn, are built around processes of transmission and processing of information. It follows then that when analyzing life activity, information flows are to be taken into account, apart from the material and energy flows. It is precisely the information flows that provide for the integrity of an organism, organizing the material and energy processes into an appropriate pattern. The need to study the laws behind the transmission and transformation of information in the organism, the importance of control processes, have been recognized comparatively recently. The works by I. M. Sechenov, I. P. Pavlov, W. Cannon and others in biology and medicine, by C. Shannon, N. Wiener and others in the field of mathematical control and communication theory, have resulted in the emergence, at long last, of the science of cybernetics which unambiguously stipulates the importance of information processes.

So, studying the processes of life activity of an integral organism, we must take into consideration the interrelationship and interaction of three basic components: material, energy and information flows.

Transmission and processing of information are no less important for understanding the mechanism of life activity than the exchange of matter and energy, for it is the former that provide for the integrity of the organism, in fact for its very existence.

This paper deals with the informational pattern of a control system which governs the physiological medium of an integral organism, the physiological medium implying herein the totality of the organs and tissues providing for the inflow of energy and matter required to sustain the organism. The physiological medium of a complex multicellular organism (that of mammals and, in particular, of human beings) is largely independent of the environment. It is precisely this property that enables a living organism to follow rather complex behaviour patterns in a changing environment. At the same time, this degree of independence of the physiological medium can be attained, as noted above, only through such a mode of functioning of the control system that would provide for the stability of the material and energy processes in the physiological medium relative to extraneous disturbances. The control system of an integral organism, just as any other control system, functions on the basis of information processing. The principles of organization of the latter will be herein referred to as the informational pattern of the control system.

Before dealing with the organization and functioning of the physiological medium control system, we shall dwell in brief on the physiological medium as an object of control.

1. Physiological medium as an object of control

The physiological medium is responsible for two major aspects of life activity: a) generation of energy required for growth, for the maintenance of tonicity and for active behaviour; and b) self-reproduction of cells and tissues without which living for any length of time is impossible. Both these processes are complicated complexes of biochemical reactions and biophysical processes. A normal course of reproduction of energy and self-reproduction of cells and tissues requires a constant supply of nutrients and oxygen and a constant elimination of exchange products. In itself, the structure of interrelationships of the material and energy flows in the organism is extremely complicated.

With a view to revealing the chief peculiarities of the physiological medium as an object of control, let us dwell in brief on the subsystem which provides for energy generation.

Simplifying the real situation somewhat, we can take that this subsystem comprises the following organs: the lungs, the liver, the intestines and the kidneys; a number of endocrine glands (hypophysis, thyroid and pancreas); the muscles (as energy and oxygen consumers); and the circulatory system whereby substances are transferred among the listed organs. The subsystem in question carries the processes of gas exchange (oxygen-carbon dioxide), carbohydrate exchange (glucose, glycogen) and energy exchange (including heat exchange).

Even in the absence of a special control system, the totality of the above processes of material and energy exchange features certain stabilizing properties. Thus, accumulation of carbohydrates in the liver tends to mitigate glucose level fluctuations in the blood (the so-called homeostatic mechanism of the liver). Such stabilizing properties manifest themselves in a system where information flows seem at first sight to be completely lacking. Close scrutiny reveals, however, that they result from a specific pattern of this system, from its structure which cannot be formed without information. So, here, too, the material and energy processes must be viewed in terms of their spatial-temporal organization, which is to say the information built into the structure of the system.

The physiological medium exhibits its intrinsic stabilizing properties only when the disturbances are small enough; these properties alone would fail to provide for life activity. However, the processes discussed proceed in the organism with a direct participation of the hormonal system which exerts a stabilizing and organizing influence on the exchange of substances and energy. The hormonal system is in effect part of the control system (the hormones involved in the workings of the energy subsystem are thyroxin, insulin and glucagon as well as the thyreotropic and pancreotropic hormones). It forms a number of closed regulation circuits having certain specific features. Thus, among other things, they are characterized by the lack of clearly defined information transmission and transformation channels (for instance, the measuring organ – receptor – can be hardly defined in the glucose-insulin circuit). It will be noted that the described part of the hormonal system is essential to the functioning of the physiological medium, which is why the concept of "energy subsystem" is to be regarded hereinafter as including part of the hormonal system.

In one of our earlier papers [2] it has been shown that, given some simplifying assumptions, the energy subsystem can be described by a set of conventional non-linear differential equations:

$$\frac{dx}{dt} = f(x, u, z),$$

where x is the vector of the system state coordinates which include the levels of the above-listed substances and the temperatures in the organs making up the subsystem as well as the temperature in the circulatory system;

u is the vector of the control responses, i.e. those parameters of the subsystem which can be varied by the organism control system; and

z is the vector of the disturbances acting on the energy subsystem (including, in particular, the energy consumed by the organism to do manual work).

At fixed values of control responses and disturbances and with the conditions of external balance met (i.e. the amount of carbohydrates or their energy equivalents introduced into the organism being equal to the energy consumed), the energy subsystem is stable, which state is achieved, as noted above, by the functioning of the hormonal system. At the same time, acted upon by disturbances, the system state coordinates are likely to pass beyond the permissible region; besides, the demand for energy is likely not to be satisfied. Then a control system is needed to provide for the integrity of the physiological medium in the presence of disturbances.

2. The informational structure of the physiological medium control system

Let us consider the space of physiological medium states and delineate within it a permissible region, that is to say a region of such values of the physiological medium parameters which allow of life activity. If a demand for a certain amount of energy is imposed on the system, then within the permissible region of states of the physiological medium a subregion will be delineated which comprises the state at which the required amount of energy can be generated. However, these states are not equivalent either, and consequently, only one of them is optimal from the viewpoint of some criterion (for example, the minimal energy or stored carbohydrate consumption criterion). It should be noted that the energy requirements, just as the optimality criterion, depend on the environment.

Thus, the process of controlling a physiological medium includes the following tasks:

1. determination of the optimality criterion on the basis of the external conditions and the state of the organism;
2. determination of the optimum state of the internal medium;
3. maintaining the physiological medium parameters at the required level.

It should also be noted that each of the tasks listed is to be accomplished in its own time scale. Thus, the physiological medium state parameters are to be stabilized against the effect of rather quick-acting disturbances; therefore essential to this process are the dynamic properties of the organs making up the energy subsystem. The determinants for the optimal state of the physiological medium are the mean values of the physiological parameters; consequently, this task may be regarded as one of static optimization taken in a time scale which is slower than the one used in the previous example.

With this in view and taking into account the complexity of the task and the need to decompose it, a hypothesis has been put forward as to the hierarchical pattern of the physiological medium control system [1]. Such a hierarchical structure comprises three levels.

The first, lower, level—homeostasis—comprises local systems which in totality maintain a certain state of the internal medium. The required values of the parameters are assigned to the homeostasis level from the higher-lying second level.

The second level of the hierarchical system provides for the optimal condition of the physiological medium corresponding to the optimum of a certain criterion, which criterion is in turn determined in the upper third level. In the course of its functioning the second level determines the values of the physiological medium parameters to be maintained by the homeostasis level.

Finally, the highest third level of the control system works out the general criterion of optimality of the physiological medium condition.

The system organized in this manner is on the whole centralized, though each of its levels has a measure of independence. Only part of the information about the state of the physiological medium reaches the highest control levels, the remaining data being utilized at the lower levels. Such a hierarchical pattern of the control system has both advantages and disadvantages. Its flexibility and reliability are patently obvious. Another important point is that the higher levels receive only specially averaged information, which simplifies their own structure. At the same time, such a system is rather slow and is likely to prove inefficient in cases where the environment changes fast enough (danger situations). Thereby, a question arises as to the possibility of combining centralized control, the upper level directly controlling the physiological medium, with the above hierarchical principle of control (it may be called decompositional as well).

The control system of an integral organism provides for the possibility of direct intervention of the higher levels in the processes proceeding in the physiological medium. This phenomenon is known as an "adaptation syndrome" arising under stress [4].

Characteristically, though, this kind of intervention does not involve a detailed restructuring of the entire physiological medium, i.e. it is not specific. The adaptation syndrome manifests itself in a release into the blood stream of a range of hormones which, in the case of the energy subsystem for example, build up the amount of glucose in the blood (adrenalin); boost the circulation function, etc. The physiological medium is thereby switched to a state where the maximum possible energy demand can be met. In such a case, naturally, there can be no question about the optimality of the second-level criterion.

Interventions of this kind are rare occasions observed only in emergencies, the rest of the time the above-described hierarchical system holding sway.

It should be additionally noted that the physiological medium may be regarded as a complex of interrelated (and relatively independent) objects. Then the second-level control may be decentralized (in other words, the general criterion of optimality is broken down according to a special pattern into particular subsystem criteria which are the ones that are optimized).

So, the informational pattern of a hierarchical control system of the physiological medium of an integral organism is characterized by the decomposition of control in time and among the objects of the physiological medium which is viewed as a complex. It is also characterized by a certain relationship of the centralized and decompositional control principles, whereby the centralized reaction is non-specific and is "invoked" only in emergencies.

The above principles of a control system can be applied to rather complicated complexes, such as those encountered in chemical engineering or economics.

In conclusion, it will be mentioned once again that unless information flows are considered alongside material and energy flows, the processes occurring in an integral organism, just as in other similar systems, can never be adequately understood.

References

[1] С. Н. Брайнес, В. Б. Свечинский: Проблемы нейрокибернетики и нейробионики (S. N. Braines, V. B. Svetchinsky: Problems of Neurocybernetics and Neurobionics). Медицина, Москва 1968.

[2] С. Н. Брайнес, В. Б. Свечинский: Структура математического описания энергетической подсистемы организма как объекта управления (S. N. Braines, V. B. Svetchinsky: The Structure of a Mathematical Description of the Energy Subsystem of the Organism as an Object of Control). Всесоюзное совещание по проблемам управления — рефераты докладов. Наука, Москва 1971.

[3] S. Braines, V. Swetschinski: Die Steuerung im Organismus. Ideen des exakten Wissens. Berlin 1969.

[4] H. Selye: The Stress of Life. McGraw-Hill, New York 1956.

INFORMATION AND MENTAL PROCESSES

Michal Stríženec

The information theory and other cybernetic disciplines bring along a novel approach to the investigation and explanation of psychological problems. The informational aspect represents a considerable contribution to a quantification of the more complex forms of mental activity. It derives from the simple fact that information intake and processing enable the organism to adjust itself to its environment. A repression of the interaction between the organism and the environment results in an enhanced entropy (for instance, forgetting). Thinking, creative activity, on the other hand, are proofs of a decrease in entropy in mental processes.

The first to point out the importance of information theory for psychology were G. A. Miller and F. C. Frick in 1949. Later, several authors came to be engaged on this problem and Frank [2] proposed the term information psychology. The relevant aspects involved in these questions have so far been most processed by Klix et al. [4; 5]. In our monograph [10] we presented a review of past researches and the results of our own experiments in this area (it also contains an extensive list of references to literature that will be of help to those interested in the problem).

Efforts made thus far to apply information theory in psychology have been aimed principally at perception, memory and learning, thinking. Here, we shall endeavour to point out in brief the principal concepts that have been derived from these studies.

Perception

Perception involves awareness of object or relations with the aid of sensory processes. Perception is considered to be a relatively independent process having its regulation and specific product in the form of an overall picture of the stimulus.

The concept of information is also utilized in explaining both the neurophysiological substrate of perception, and the phenomena of perception as such. Thus, for instance, a comparison is made of the theoretical information capacity of various sensory organs (vision –

3 million bits, hearing – 8 000 bits), the speed of information transfer in man (in particular the dependence of the response time on the quantity of information input in man). Perception reflects the probability structure of the external reality, and hence, an investigation was made, for instance, of the a priori probabilistic hypotheses in relation to the recognition of various properties of the objects being perceived. It was found that the utilization of information depends on an optimum sequence of signal presentation. A study of redundancy in visual stimulation showed information to concentrate, for example, along the outlines of a figure. Greene and Courtis [3] have however provided experimental evidence that the quantity of information is a function of the perceiving subject and not one of the stimulus source. In their view, conditions for utilizing the information theory in visual perception have not been met.

It is becoming more and more evident that when the informational value of any event is to be determined, its subjective probability must be taken into consideration. Definitions of subjective probability are far from uniform. Similarly in its measurement one also proceeds, for example, with the aid of prediction or expectation judgments on independence of various classes of events. Difficulties in the measurement of subjective probabilities derive from the fact that here a complex of factors has to be taken into consideration, factors the objective probability of which is not known. In addition, the effect of age, experience, number of available alternatives also plays a role here. However, the help of a direct psychological judgment, or Edward's decision models, the problem of quantification of subjective probabilities may be seen in utilizing Bayes' theorem which enables to express the change of subjective probabilities of hypotheses.

Now, we shall mention some of the conclusions from our studies in this area. In our first researches we investigated the effect of the quantity of irrelevant information (and other factors) on disjunctive reaction time. The quantity of transferred information was found to increase linearly with an increase in the amount of presented irrelevant information. In addition, it also depends on the position of the signal in a sequence of signals and their spatial distribution.

In a further work we showed that the information transfer increases with a maximum stimulus-response compatibility. With the aid a modification of the method according to Attneave and Arnoult and Bayes' theorem, we determined the subjective probability of occurrence of various letters of the alphabet. It was found that the searching process of critical points is not linear, and an optimum strategy occurred only rarely.

We also investigated [9] the threshold of chance refusal with the method according to Alberoni, where we compared various age and sex groups. In investigating the possibilities of using psychological scales to determine subjective probability, we made use of an adjusted list of 10 events according to Galanter (in various age groups), then we followed the relationship between subjective probability of success and the number of hypothetical trials (where our results proved similar to those of Cohen).

In conclusion it may be stated that attempts thus far at utilizing the concepts of contemporary information theory in research on perception are justified solely in isolated simple components of this complex process. Research on response time points to the necessity of further elaboration of the mathematical apparatus of information theory. Characteristic for human perception is an active-selective orientation. It results from an interaction of stimulation and set. Set may derive from learned probabilistic relationships. A mechanical judgment of the stimulus situation in bit/sec is unsuitable in an analysis of perceptual processes. According to Fatkin [1], the notions of the probability theory, the theory of games and decision-making should be applied here. It should not be forgotten that a sense-perceived image captures the temporal-spatial organization of reality, preceding experience, and social influences. Quantitative models of human perception must make account of these facts, i.e. must start from a real course of the relevant process. From the aspect of contribution to the psychological theory, we consider as the most significant the probability principle found in perceptual processes. The quantity of information presented should be judged against the background of subjective probability of signal occurrence, which is in turn affected by experience and similar phenomena.

Memory and Learning

As a rule, a distinction is made between immediate, short-term and long-term memory. The effect of the information theory has here been manifest mainly in a novel quantitative approach to the evaluation of memory-stored or recalled material. Data on the capacity of human memory are in the range 10^9 to 10^{13} bits. The rate of information inflow is about 0.5 bit per sec. The limited capacity of immediate memory may be improved by organizing the stimuli into "clusters." These groupings are formed during the learning process. In this connection, the so-called magic number 7 is spoken on (7 ± 2 stimuli). It has been found that in learning by heart, achievement depends on the length of sequence and not on the quantity of embodied information. During the memorizing process, information is processed both in that information in symbols declines (the symbol alphabet becomes limited) and that the number of symbols diminishes (transition to a new alphabet). Disintegration of the memory trace depends on the original redundance, on coding and on the importance of information in other traces.

Linhart [6] understands learning as a self-regulating process and in this connetion also takes note of verbal information, information selection, stages of information processing during the course of learning. Several authors have investigated the role of feedback in learning (stimulating and orienting properties of feedback). It appears that presentation of an adequate quantity of information in a suitable sensory way and immediately after performance significantly accelerates the learning process. For reason of space, the extensive area of information transfer in learning and programmed teaching cannot be dealt with here in any detail. It may be said in conclusion that an informational evaluation of memory capacity and a cybernetic modelling of learning have contributed to revive interest in these processes and enabled a more effective cooperation between neurophysiology and psychology in elucidating the learning mechanisms. The information theory has contributed in particular to a novel view on the quantification of the stimulus and performance component in memory (questions of coding, clustering, effect of redundancy).

Thinking

The information theory can contribute to an optimizing of successive reduction of uncertainty in problem solving. This involves a probability of alternatives, their informational contents, relative redundancy. According to Oerter [8] redundancy enables us to differentiate between productive and reproductive thinking (less redundant steps in the initial stage of problem solving point to a higher productiveness). Following the practical determinations of the informational content of a problem, difficulties may arise (especially in alternatives with a different probability of occurrence, and also because man need not be familiar with the objective probability of occurrence of alternatives).

Of questions relating to thinking, a study has been made for instance, of the probabilistic character of the structure of a task, duration of problem solving, algorithms of mental activity, solving a logical task with uncertainty. The extensive area of heuristic programming and artificial intelligence is beyond the scope of this collection and therefore will not be dealt with here. Efforts at capturing the complex nature of human thinking have been made by Newell [7] in models of information processing. However, no account is here made of the specificity of human thinking-purposefulness, selectivity. Man finds the manner of solving a task during the course of problem solving. In heuristic problems, this maner is included already at the start as "known".

We have investigated the sequential character of thinking with the aid of the "PSI" apparatus, where the problem involved 9 bits. Is was found that unsuccessful performance embodieed a significantly longer duration of the test, more repetitions, but also a higher quantity of information obtained. Unsuccessful Ss began close to the target and only then did they proceed to the initial questions. For more detailed data on the research see Stríženec [11].

Of all the manners of applications used up to now, it is heuristic programming that has contributed in a large measure to a thorough analysis of the structure of thinking processes.

The contribution on the information theory to psychology resides principally in its unifying methodological aspect and its pointing to certain relatively neglected areas of psychological research. Although quantification in processing results of empirical researches has long been currently used in psychology, information measures represent an attempt at a uniform measurement of the input and the output component of man's behaviour. A not less significant aspect here is also the novel theoretical platform for cooperation between psychology, neurophysiology and certain technico-economic disciplines. However, a mutual enrichment of psychology and the cybernetic branches will proceed successfully only if a real course of mental processes is taken as the starting point of the studies.

References

[1] Л. В. Фаткин: Статистический анализ процессов восприятия (L. V. Fatkin: Statistical analysis of perception processes). In: Инженерная псыхология (А. Н.Леонтев). Изд. Москов. униврс., Москва 1964, 42—51.

[2] H. Frank: Kybernetische Grundlagen der Pädagogik. Agis-Verlag, Baden-Baden 1962.
[3] R. T. Green, M. C. Courtis: Information theory and figure perception: The metaphor that failed. Acta Psychologica *30* (1966), 1, 12—36.
[4] F. Klix (Hrsg.): Kybernetische Analysen geistiger Prozesse. DVW, Berlin 1968.
[5] F. Klix: Information und Verhalten. DVW, Berlin 1971.
[6] J. Linhart: Psychologické problémy teorie učení. Academia, Praha 1965.
[7] A. Newell, H. A. Simon: Human problem solving. Prentice Hall, New Jersey 1972.
[8] R. Oerter: Psychologie des Denkens. L. Auer, Donauwörth 1971.
[9] M. Strížеnec: Zur experimentellen Bestimmung der subjektiven Wahrscheinlichkeit. Ztsch. f. Psychologie *171* (1965), 377—382.
[10] M. Strížеnec: Psychológia a kybernetika. VSAV, Bratislava 1966.
[11] M. Strížеnec: Quelques aspects méthodologiques dans la recherche sur la pensée de l'opérateur. Travail humain *31* (1968), 3—4, 269—282.

A CONTRIBUTION TO INFORMATION ANALYSIS OF IDENTIFICATION PROCESSES

J. Linhart, S. Vohník, P. Horák

Contemporary psychological researches on thinking and learning are increasingly searching for an exact microanalysis of cognitive processes, emanating from modern logic.

An empirical analysis of hypotheses and strategies, e.g. in the descriptive form suggested by Bruner et al. [2], has not resulted in any substantial progress in the psychology of concept formation and learning. Piaget [18] pointed out that strategies as conceived by Bruner are left isolated and not coordinated in structures, without which the development of cognitive processes as the undivided activity of a subject cannot be understood. The necessity of studying thinking as an objective process and cognitive selforganization was stressed by S. L. Rubinstein [21]. His work, in particular that concerned with information analysis, was followed by V. N. Pushkin [20], Pospelov and Pushkin [19], J. Linhart [11; 12] and others. At the same time, the researches into problem solving carried out by Newell and Simon [16] with the aid of information analysis of records indicate the necessity of a deeper elaboration of information *microanalysis of cognitive activity and thought processes*. The dialectic of cognition, in fact, implies that the understanding of integrity can be achieved through a microanalysis of systems under investigation.

1. The basic concepts in cognitive process

There exist many experiments aiming at the analysis of concept cognition in terms of the *information theory*. The application of this theory in psychology is not easy and gives rise to a number of questions (A. Quastler [17], M. Stríženec [23]). At the present time, and particularly in connection with the study of heuristic programmes, *information analysis* is being elaborated (Newell and Simon [16]). In its application in the sphere of sign concept identification and concept formation we proceed from the following methodological assumptions: in identification – or, more generally, in problem solving – the subject being in the state of initial uncertainty (being unacquainted with the situation) can choose relevant or irrelevant signs with the same probability, which brings him varying amounts of information; the difficulties which the subject experiences in concept formation and problem solving depend also on (e.g. in the problem structure and in the relations among its items there can be key points whose solution brings the greatest amount of information); in information analysis we investigate further how much information is supplied by positive and negative instances, i.e. by knowledge that a definite stimulus is or is not an instance of the pursued concept.

The cognitive process does not arise where either all information is given at the beginning, or none is given; further, it cannot arise where there is no connection between the known and the unknown – on the contrary, it arises where an objective relation between the known and the unknown exists; the cognizing subject does not create these relations by himself, the conditions for their occurrence exist both in the object itself and in the relation between the subject and object; the cognising subject, however, must make them clear. Accordingly, in a study of cognitive process the following two relations must be followed: the relation subject-object and the relation explicit-implicit.

A specific feature of *identification* is the presumption of discrimination and generalization of *meanings*. A complex, hitherto unclear process is involved, in which the subject transforms reflected signs (Merkmale) of things (their qualities, relations) into signs (Zeichen) becoming carriers of meanings.

At a formal level the relation explicit-implicit can be expressed by means of the relation extensional-intensional. This conception requires the elaboration of the category of identification and classification in relation to meaning, generalization and abstraction. It follows from our finding (J. Linhart [12]) that the subject does not discriminate and generalize objects and stimuli themselves but the signs and meanings singled out from them. Therefore, in a study of identification processes the primary question is the *genesis of image and of meaning*. From psychological research it follows that the genesis of the structure sign-meaning is connected with processes of *generalization* and *abstraction* [24; 21; 12].

We consider *generalization* to be a process taking place on different levels:

1. On the level of irradiation of excitement and inhibition

a) it is going on in every direction at random;

b) it is going on in selected directions along association trails.

2. On the level of signs and meanings

a) it is bringing distant phenomenon of reality nearer;

b) generalization makes it possible to operate with forms of cognition (e.g. logical formulae);

c) generalization gives the effect of "immediateness" of cognition ("insight").

3. Sign-meaning generalization is further divided into perceptual and thought. *Thought generalization* is further divided according to the relation between analysis and synthesis:

a) firstly, there is the empirical, i.e. gradual (inductive one; the analysis of and search for identical signs and meanings occur here regularly in a hypothetic way on the basis of feedback verification); here analysis precedes and facilitates the rise of generalization;

b) secondly, there is the theoretical one where generalization has been formed earlier and where immediate generalization in the form of thought insight occurs; here analysis arises and develops following from generalization previously achieved.

Abstraction is taken as a process going on from one level (the lower one) to a higher one in hierarchic pattern (organization, structure) of cognitive processes. Abstraction is a process taking place on the sign-concept level. On the lower, directly perceptual level, a selective perceptual discrimination corresponds to abstraction.

The above mentioned sign-meaning level is connected with the role of verbal symbols in problem solving (including identifications tasks). In the sense of dialectics of cognition it can be said that every single word is general and "that every word generalizes" [9]. *Man separates definite signs from sensual material and accentuates them as a common feature or relevant sign of many concepts and of a class of phenomena.* A process of *abstraction* is involved which, however, cannot be described according to the assumptions of philosophical empirism, as a mere singling out of general features; in particular, a process is involved in which in a specific way the shaping of *sign-meaning structures* and the corresponding *information invariants* participate. The process of categorization is connected with it as well; it enables grouping and classifying things on the basis of their common qualities and relevant relations.

We distinguish abstraction according to relations between analysis and synthesis and according to relation to operations in the two following kinds:

a) empirical abstraction ensuring from the separation of relevant sign-meanings by gradual analysis (the inductive way);

b) reflexive abstraction ensuring from thought operations.

Identification is conceived as a choice of a sign-meaning as relevant for a definite object or problem situation; the choice of a relevant sign would be at the same time the choice of a relevant (suitable etc.) object. The establishing of a relevant meaning is connected with concept formation. What is to be identified can be given in advance (e.g. by means of a pattern) or is to be found in the course of the process of cognition and solution. In solving identification tasks the human brain works with incomplete information, uses analogies, guesses, probability evaluation, removes uncertainty and superfluity. Accordingly, we consider discrimination of signs both relevant and irrelevant to be the starting point in the information analysis of identification and concept-formation processes. Understanding a sign of a cognised object and assigning meaning to the sign is a condition of idetification. As mentioned above, identification in this sense means discrimination of relevant signs from irrelevant ones (in dependence on the choice of a viewpoint) and generalization of the relevant meaning. In information analysis we investigate further how much information is supplied by positive and negative instances, i.e. by knowledge that a definite stimulus is or is not an instance of the pursued concept. The choice of relevance is generally connected with adopting a given subjective criterion in cognising an object.

In order that the processs of learning and concept formation may take place a variety delimitation must exist in the problem situation (W. R. Ashby [13]). An arrangement of signs as relevant and irrelevant which is not known to the solver is one form of variety delimitation. The identification of the relevant sign is based on the cognition of this arrangement, on the reconstruction of the initial situation.

Classification (categorization) is connected with identification. With respect to classification, generalization, abstraction and identification are basic processes and operations. The role of classification in cognition is as follows:

a) finding a common class for certain phenomena and finding that an object belongs to a class (that the object is an element of the class); in this sense classification means finding the way of identifying objects, dealing with discriminative objects as equivalent ones;

b) classification is the way in which man finds variety delimitation, i.e. achieves a necessary reduction of a complex situation by means of common signs; this is a condition of creating an appropriate effective system and an adequate system of operations;

c) with regard to activity, classification widens the ability of the subject to anticipate future actions and organize his behaviour in accordance with a goal;

d) classification is effected (at least in the situation chosen by us) by assigning a meaning to the common sign, which becomes a guide to the operations performed;

e) this implies that identification and classification are conditions of the formation of concepts, their systems and relations between classes;

The processes of generalization, abstraction, identification and classification are conditions of the genesis of the concept. The last step to the concept is a stabilized meaning whose content is that which is equal (similar) in the objects and common to them. The formation of a stabilized meaning—as an elementary concept—necessitates the singling out of the relevant sign, common to a set of elements in the objects cognised and compared.

As a starting point in the information analysis of concept-forming procesess we regard the division of signs into *relevant and irrelevant ones.* From the point of view of psychology the problem of relevant and irrelevant signs links up with the study of semiotic processes, particularly the relations between sign, meaning and sense. The relevance of a sign is directly dependent on its meaning. As the meaning and the sense of a sign are conditioned by the structural relations (it being possible to understand sense and meaning as the centre of relations in the cognitive focus), in the genesis of the relevance of the sign it is necessary to take into account the relation association bonds (see [3]). Information analysis further points out that there exist relations between relevance and irrelevance of signs and stimulus superfluity in concept formation; (the redundant) superfluous signs need not be considered to be important and relevant. Relevant signs are those which bring *useful information* with respect to solving a given task. In the theory of knowledge the question of relevance and irrelevance of a sign is connected with that of the transition from the cognition of random phenomena to grasping the substance. The problem involved in the theory of knowledge, until recently investigated mostly in terms of philosophy, now rests upon a widely elaborated experimental basis.

In task solving the human brain works with incomplete information, making use of analogies, estimates, probability evaluation, removing uncertainty and redundance. The result of the process in the shaping of the meaning as *an invariant of information*, i.e. that which remains stable in the course of the transformation and information contained in it [15; 13]. The meanings of signs are a basis for the formation of concepts and their use in social practice.

Meanings as well as concepts are not substances independent of the subject,[1]) but are reflections of reality and at the same time semiotic functions. In this *structural and functional conception*, the function of meanings is carried out by signs, objects, processes and states. In concept formation it is the structural aspect that is important for on its basis W. R. Gardner [4] distinguishes between paired associate learning and concept learning. In the former a single response is associated with each stimulus; contrarily in the latter a greater quantity of stimuli is assigned to one and the same response. In other words, in his response the subject finds and expresses what is common to more or at least two stimuli. Considering only the quantitative aspect we can say that in concept learning there is a grater quantity of stimuli than responses. This concept redundance implies that some stimuli, or their signs, are irrelevant. For example, in Bruner's investigation for the concept of "red circle" those attributes are irrelevant which express their number, circle frames and the like; knowing them reinforces the connection of relevant signs with the correct response. The knowledge of the fact that some signs are not essential for a given concept or for solving a given problem will enable the solver to find the relevant relations more easily and thus to reach the required concept or principle of solving. Ascertaining the sign relevance is also frequent in *inductive* reasoning, in seeking causal relationships between two phenomena, in medical and psychological diagnostics, etc.

The concept consists in singling out a group of cognized objects according to one general relevant sign-meaning by way of isolating abstraction; the concept comes into existence when a number of abstracted signs-meanings are synthetized. We regard abstracting synthesis as the basic form of *concept thinking*, a decisive role being played by the word as the main device of abstracting identification and theoretical generalization.

In concept identification processes inductive-probabilistic components are combined with logical-deductive ones. Roughly speaking, concept formation presents the following relations between the two components: a) at the initial stage of solving, inductive-probabilistic processes prevail, while their role gradually declines (the first phase); b) in the second phase, in connection with sufficient probability the choice of a given hypothesis takes place, while subjective uncertainty speedily decreases; c) in the third phase, deduction gains ground. The relations betwece, inductive and deductive components vary in dependence on previous experienen on the distance from the goal, on the arrangement of the series of tasks and on other conditions.

In the genesis of the concept – in the course of solving and learning – the subject ensues from attributes of things to shape a universe of objects which he can identify as elements of the class sought. There are two possible situations: a) either *concept attainment* where the subject attains and identifies a concept previously defined, b) or *concept formation.* In the latter case the concept is sought in an unknown set of objects – here concept learning is in harmony with problem solving. The models of concept formation are concerned with the question as to how the subject discovers rules unfamiliar to him for arranging experience, groups of symptoms and classes of objects. An explanation of these processes on the basis of experimen-

[1]) Concepts are regarded as substances of things by objective idealism (neo-Hegelianism, neo-Platonism etc.).

tal research is to permit a deeper understanding of the processes of the formation of new concepts.[2])

From concept formation, where the subject himself on the basis of his own activity in problem solving arrives at concept generalization we distinguish concept attainment, where the subject is to fix in his mind and memory ready-made concepts.

We distinguish concepts on two different levels: a) empirical concepts and b) logical-mathematical concepts, coming into being in the process of reflective abstraction.

In the formation of the concept as a meaning of a given name, it is necessary to define its *designate and its range.* Both cases involve the knowledge as to whether a given phenomenon is or is not an instance (example) of a given concept. In psychology the so-called positive and negative instances of the concept are referred to in this case. In the conclusions of older researches it was alleged (Smoke [22] and others) that the subject gains little profit from negative instances; according to these researches the information that a given phenomenon does not belongs to a given concept, contributes little to the solution of a given problem. This point of view was recognized as generally valid for a long time. New and more exact research methods have shown that such a conclusion is premature. In Smoke's experimental arrangement there was a priori a different quantity of information for positive and negative choices; the negative choices were bringing less information because they did not eliminate so many possibilities (hypotheses) as positive examples of the concept. In other words, the acquired quantity of information is dependent on the structure of the problem, in which, as mentioned above, the quantity of information which the subject is to acquire is distributed in an uneven way. Some positive or, contrarily, negative choices can bring more information than other choices. This new point of view was confirmed by the research of Hovland and Weiss [6] and, more recently, also by that of J. Linhart [12]. In their arrangement the positive and negative examples of the concept sought transmitted approximately the same quantity of information. According to the information theory, from which the authors proceed, it is possible to vary the nature of the problem so that the quantity of information in positive and negative instances may be mutually varied. For example, it is possible to create a problem which can be logically solved by four positive or four negative instances.

The concept formation in the *process of problem solving is characterized by the character of a functional exploitation of the rising concept in various phases in thought activity.* In the first phase it is a process of sizing up and finding the relevance of the sign and its meaning, in the second phase it is the generalization of the concept. From the initial data, items and signs, contained in the problem situation, the subject involved in concept formation infers that there

[2]) The knowledge of the processes of concept formation is important both for education and instruction and in the construction of programmes for self-organization systems, automata and the like. Bruner et al. [2] endeavoured to eliminate terminological confusion in the following way: they drew a line between "concept attainment" and "concept formation". In their study they say that the component that they called "concept attainment" unlike "concept formation" is seeking and verifying attributes, which can distinguish what is relevant from what is not relevant in various categories; it is the seeking of adequate and valid anticipation cues. The process of *concept formation* (or in other words the process of category formation) is according to Bruner a primary process, it is the first step towards concept attainment.

are relations between them. Generally speaking, the problem is implicitly given as a system of signs (symptoms, attributes) and their relations, which the solver is to cognise, group and classify. The situation analysis and finding relevant signs and their meanings and their verbal designation are followed by the phase of generalization bearing on new objects and on the solution of new situations.

2. Some examples of identification and classification

Example 1: On a desk two sets of books are placed – one containing only Spanish books, the other only Italian books. The subject is to determine the classifying signs so that he may assign the next book (Spanish or Italian) to the correct set. If we choose the experimental subjects at random from the given population, it is necessary to take into consideration two things:

a) The experimental subject belongs to the first group of persons who do not know either Spanish or Italian.

b) The experimental subject belongs to the second group of persons who know at least one of the languages in question or, as the case may be, both. A task of this type does not represent any process of cognition for the subject belonging to the second group; in the next book he will ascertain the language in which it is written and, accordingly, he adds it to one or the other set. The task becomes a complicated one for an experimental subject belonging to the first group. First he focusses his attention on some outer signs (type of binding, colour of cover, type of paper, etc.) and only when he finds that these outer signs do not establish a classification key, he either gives up solving the problem or he detects in both groups language differences and solves the task. Solving the task, therefore, depends on defining new signs-meanings.

Example 2: *A successive concept identification on ellipses with dots.* In an experimental investigation with rows presented successively, the subjects were to learn how to distinguish two signs: namely to tell a circle with dots outside the ellipse from a circle with dots inside the ellipse (see Fig. 1).[3]) The figures were projected according to a diagram previously defined by the experimentor (see Table 1). In one half of the ellipses the dots are inside the ellipses, while in the second half (i.e. 15) the dots are outside. Identification is to be effected by the subject in the following way: if the dots are inside the ellipses he is to press the button, in other cases, i.e. when the dots are outside the ellipses, the button is not to be pressed. In the instructions

[3]) It is an example taken from an experimental investigation of sign concept identification carried out by K. Tříska and J. Linhart (see J. Linhart [16]). The investigation concerns the identification of artificial concepts on the basis of recognition of two relevant signs. The representative figure of the ellipses is one of the most adequate figures for the identification of a relevant sign. The figures were presented simultaneously and successively.

the subject is told that he is to learn to press the button, if he does so correctly a white light will appear and if he does so incorrectly a red light will appear. If the subject does not press the button, he receives no feedback, i.e. no light will appear.

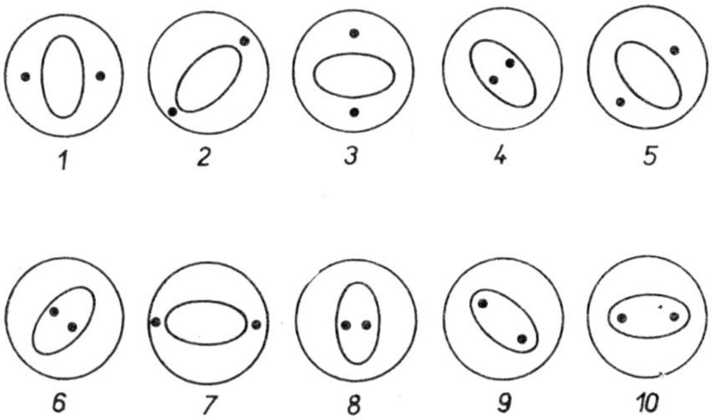

Fig. 1. Set of figures designed for projection.

TABLE: Schedule of Projection of Figures

Serial number of the figure	Number of the figure projected to fig	Serial number of the figure	Number of the figure projected to fig	Serial number of the figure	Number of the figure projected to fig
1	1	11	7	21	1
2	2	12	5	22	8
3	3	13	10	23	9
4	4	14	3	24	3
5	5	15	9	25	2
6	6	16	2	26	5
7	7	17	8	27	10
8	8	18	1	28	7
9	9	19	6	29	4
10	10	20	4	30	6

This experiment is an optimum example of identification. The experimental subject is to find in the objects presented a structure, i.e. he learns to distinguish the relation "to lie outside or inside". This implies that unlike the first examination of the figure as a whole the subject – in the correct solution – he disregards the non-substantial (i.e. the irrelevant) parts (e.g. the orientation of the ellipses, the orientation of the dots, the position of the dots with respect to the axes of the ellipses, etc.). In the instructions the subject is not given any direction in solving nor the sought shape of the figure, i.e. incomplete information as to the goal and procedure is given.

3. Formal processing

It follows from the analysis that identification is the basic cognitive process. One of the ways towards its deeper comprehension is its detailed formal processing. In our formulation we shall omit the initial random phase, for it can be processed by an accessible probability apparatus. We are primarily concerned with a description (by means of logical devices) of the activity of the man who is to solve an identification task. We shall deal with a formal description of experimental results on identification (SCI[4])) and problem solving (DOMINO[5]), etc.), which from the point of view of psychology were analysed by J. Linhart [10; 12]. The advantage of experimental identification tasks, in comparison with identification in current life is that in the experiment we have simplified the structure of the problem so that it is possible to describe easily the whole universe. The experimental material, the goal of the experiment and the individual steps in solving can be described uniformly for all subjects by a finite number of predicates and constants. In a general case the situation is much more complicated (see example 1; each new subject of the first group can concentrate his attention on the next not yet examined sign, i.e. the set of predicates cannot be limited beforehand by a constant – i.e. in this case the set of predicates is the least numerous). The knowledge of the universe in this case will permit a more exact way of expressing oneself and a more concrete formalization. We shall refrain from using the distinction between syntax and semantics usual in mathematical logic. The individual steps will be illustrated by means of example 2 from the foregoing paragraph.

The description of objects with respect to which identification is carried out, can be effected by means of two finite sets:

$$\text{Pred} = \{P_1, P_2, \ldots, P_N\}; \qquad \text{Const} = \{c_1, c_2, \ldots, c_M\},$$

where Pred is a set of *predicates* and Const. is a set of *constants*. In each predicate its arity is given (i.e. the function Ar: Pred $\to \omega$; ω denotes the set of natural numbers).

What is the situation, say, in example 2 (ellipse)? The basic elements of each figure with the given set is the ellipse and the dots. As, however, the description of each figure only with the aid of these two constants would be complicated and, further, such a description would not correspond to psychological assumptions on perception (man perceives wholes), we shall use the following set of constants:

$$\text{Const} = \{c_1, c_2, c_3, c_4, c_5\},$$

where c_1 – two dots,
c_2 – horizontally oriented ellipse,
c_3 – vertically oriented ellipse,

[4]) SCI stands for sign concept identification, which is identification of concepts on the basis of their signs.

[5]) Domino game, which we also call DT(1+3), is played with four Domino stones; each stone contains two colour circles which represent signs. The subject is assigned the task of arranging stones so that neighbouring stones contain one sign in common.

c_4 – an obliquely oriented ellipse, such that its main axis forms with the horizontal straight line the angle $+45°$, the orientation being counter-clockwise,

c_5 – an obliquely oriented ellipse such that its main axis forms with the horizontal straight line the angle $-45°$, the choice of orientation being the same.

In order that the figure might be exactly described it is necessary to determine further:

1. if the dots are inside or outside the ellipse,
2. if the dots lie on the main or the subsidiary axis of the ellipse.

Accordingly, we write the set

$$\text{Pred} = \{P_1, P_2\},$$

where $P_1(c_1 e)$ – dots c_1 lie inside the ellipse e,
$P_2(c_1 e)$ – dots c_1 lie on the main axis of the ellipse e.

As we know that the case where one dot would be inside the ellipse and the other outsied cannot arise and as we know, further, that if the dots do not lie on the main axis, they necessarily lie on the subsidiary axis, we shall regard these two predicates as sufficient for a complete description of the figures, presented in the experiment. The set of all atomic Atfm_0 formulae can now be defined as follows:

$$\text{Atfm}_0 = \{P_i(t_1, \ldots, t_{\text{Ar}(P_i)});\ P_i \in \text{Pred}\ \&\ t_1, \ldots, t_{\text{Ar}(P_i)} \in \text{Const}\}.$$

Let us keep in mind that owing to the arrangement of the experiment some atomic formulae Atfm_0 do not come into consideration. For example, in our case $P_1(c_1, c_1)$ or $P_2(c_3, c_5)$ etc. cannot take place. All such inadmissible atomic formulae Atfm_0 are deleted, i.e. we shall create Atfm_1.

Atomic formulae cover individual groups of the elements of such figures that we have presented in our research (see Fig. 1 and Table 1), for instance the position of both dots at the same time, placing of the ellipse, placing the dots on the axes of the ellipse, etc. Therefore, there is the description of basic perception elements, which are a condition of any identification.

In our case we have:

$$\text{Atfm}_1 = \{P_1(c_1, c_2), P_1(c_1, c_3), P_1(c_1, c_4), P_1(c_1, c_5),$$
$$P_2(c_1, c_2), P_2(c_1, c_3), P_2(c_1, c_4), P_2(c_1, c_5)\}.$$

Further, we shall form a set of *all descriptions* (the description of each object being given by the elementary conjunction of length N):

$$\Pi_0 = \{\bigwedge_{i=1}^{N} \varepsilon_i \alpha_i;\ \alpha_i \in \text{Atfm}_1\ \&\ \varepsilon_i \in \{\neg, \Lambda\},\quad (i \neq j) \Rightarrow (\alpha_i \neq \alpha_j)\}.$$

Here also it is necessary to realize that owing to the arrangement of the experiment some elementary conjunctions do not come into consideration, so that by eliminating them we shall have from Π_0 the set Π_1. The set Π_1 is an expression of a formal description of figures, which is to be identified by the subject. In particular, it expresses the fact that the subject perceives the given figure, e.g. a circle with an ellipse and dots, as a *whole*. The whole perceived in this way is distinguished by the subject from the other figures because he is able to single out relevant signs, which are carriers of meanings permitting a specification of a given class of figures. In our case, for instance, we cannot have

$$P_1(c_1, c_5) \mathbin{\&} P_1(c_1, c_3) \quad \text{or} \quad P_1(c_1, c_5) \mathbin{\&} P_2(c_1, c_3).$$

Such formula will not express for us the description of the figure presented.
Thus in our case:

$$\Pi_1 = \{\varepsilon_1 P_1(c_1, \gamma) \mathbin{\&} \varepsilon_2 P_2(c_1, \gamma);\, \varepsilon_1, \varepsilon_2 \in \{\neg, \Lambda\}, \gamma \in \{c_2, c_3, c_4, c_5\}\}.$$

or

$$\Pi_1 = \{P_1(c_1, c_2) \mathbin{\&} P_2(c_1, c_2), F_1(c_1, c_2) \mathbin{\&} \neg P_2(c_1, c_2), \neg P_1(c_1, c_2) \mathbin{\&} P_2(c_1, c_2)$$
$$\neg P_1(c_1, c_2) \mathbin{\&} \neg P_2(c_1, c_2), \ldots, \neg P_1(c_1, c_5) \mathbin{\&} P_2(c_1, c_5), \neg P_1(c_1, c_5) \mathbin{\&}$$
$$\mathbin{\&} \neg P_2(c_1, c_5)\}.$$

From now onwards each object can be regarded as an element of the set Π_1. In terms of mathematical logic the description is a closed formula without quantificator. During the experiment the subject is given a sequence of figures; accordingly, this can be expressed as follows:

$$\{\pi_n\}_{n \in \omega}; \quad \pi_n \in \Pi_1.$$

If the number of figures presented is always the same, the sequence is finite, if the number of figures presented depends on a successful solving by subjects, the sequence can be regarded as infinite.

Let us now pass on to a very important turn taking place in identification. From previous presentations or under another influence the subject will notice a part of the object presented or some relation and overlooks the rest. Actually, what does the subject accomplish by doing just that? The subject effects a general and abstract mapping (GAM). He decides, indeed, that just as with the given object he would act with other objects which have this or that sign in common. Thus, the mapping of abstraction is the mapping $F\colon \Pi_1 \to \mathcal{P}(\Pi_1)$, where the symbol $\mathcal{P}(\Pi_1)$ denotes a set of all subsets of the set Π_1, $\pi \in F(\pi)$ being valid. The next requirement laid on the mapping of abstraction is given below. By way of example let us present one of the possible mapping GAM: let us assume that the subject believes that the orientation of the ellipse is of no importance. In this case the mapping of abstraction will look as follows:

$$F_1(\varepsilon_1 P_1(c_1, \gamma_1) \mathbin{\&} \varepsilon_2 P_2(c_1, \gamma_1)) = \{\varepsilon_1 P_1(c_1, \gamma) \mathbin{\&} \varepsilon_2 P_2(c_1, \gamma);\ \gamma \in \{c_2, c_3, c_4, c_5\}\}.$$

If the subject believes that no importance is attached to whether the dots are outside or inside the ellipsis, it can be described by the following mapping of abstraction:

$$F_2(\varepsilon_1 P_1(c_1, \gamma) \mathbin{\&} \varepsilon_2 P_2(c_1, \gamma)) = \{\varepsilon_1 P_1(c_1, \gamma) \mathbin{\&} \varepsilon_2 P_2(c_1, \gamma), \bar{\varepsilon}_1 P_1(c_1, \gamma) \mathbin{\&} \varepsilon_2 P_2(c_1, \gamma)\}.$$

where

$$\bar{\varepsilon} = \begin{cases} \neg & \text{if } \varepsilon = \Lambda \\ \Lambda & \text{if } \varepsilon = \neg \end{cases}$$

If the subject believes that no importance is attached to the orientation of dots with respect to the axes of the ellipsis, it can be described by the following mapping of abstraction:

$$F_3(\varepsilon_1 P_1(c_1, \gamma) \,\&\, \varepsilon_2 P_2(c_1, \gamma)) = \{\varepsilon_1 P_1(c_1, \gamma) \,\&\, \varepsilon_2 P_2(c_1, \gamma), \varepsilon_1 P_1(c_1, \gamma) \,\&\, \bar{\varepsilon}_2 P_2(c_1, \gamma)\}.$$

The correct mapping of generalization and abstraction, which the subject is to effect in the solving, is a mapping defined in the following way:

$$F_4(\pi) = F_3(F_1(\pi)).$$

The record mentioned above expresses the psychological fact that in solving the identification task analysed by us the subject takes into account only two *relevant signs* belonging to the structure of figures presented (see Fig. 1). On the basis of processes of generalization and abstraction the solver of the problem attains the discerning knowledge enabling him to see that the dots – which can be either outside or inside the ellipsis – are a relevant sign for establishing the sought class of objects.

The sets $F(\pi)$ can have a considerable number of elements. In particular where there are many constants with respect to which the subject believes that they are of no importance. At this point, introducing variables seems the obvious thing to do. Thus, we shall add the set of variables Var to the sets Pred and Const. We shall form the sets

$$\text{Atfm}' = \{P_i(t_1, \ldots, t_{\text{Ar}(P_i)}); t_1, \ldots, t_{\text{Ar}(P_i)} \in \text{Const} \cup \text{Var}\},$$

$$\Pi' = \{\bigwedge_{i=1}^{N} \varepsilon_i \alpha'_i ; \alpha'_i \in \text{Atfm}', \varepsilon_i \in \{\neg, \Lambda\}, (i \neq j) \Rightarrow (\alpha'_i \neq \alpha'_j)\}.$$

We are now defining the mapping $S: \Pi' \to \Pi_0$, which will come in useful later.

$$S(\varphi) = \{\text{Sb}_{t_1, \ldots, t_n}^{x_1, \ldots, x_n}(\varphi); t_1, \ldots, t_n \in \text{Const}, \{x_1, \ldots, x_n\} \subseteq V(\varphi)\},$$

where $\varphi \in \Pi$ and $V(\varphi)$ is a set of all variables which occur in φ. Similarly to Π_0 here in Π' we shall find formulae that owing to the organization of the experiment have no sense: by eliminating them we shall obtain Π_2.

We write

$$\Pi_2 = \{\varphi \in \Pi'; S(\varphi) \cap \Pi_1 \neq \emptyset\}.$$

For each $\pi \in \Pi_1$ we can define $\hat{\pi}$, which will be a set of those formulae from Π_2, which after the substitution of constants for the variables, will give the formulae π. Thus we can write:

$$\hat{\pi} = \{\varphi \in \Pi_2 ; \pi \in S(\varphi)\}.$$

After these adjustments we can go on to a formalization of what led to the introduction of variables. If it satisfies our intuitive assumptions, then there should exist a formula $\varphi \in \Pi_2$ such that

$$S(\varphi) \subseteq F(\pi) \qquad (\pi \in \Pi_1). \qquad (*)$$

This formula φ will certainly be an element of $\hat{\pi}$. From $\hat{\pi}$ we shall take the formula $\bar{\varphi}$, which is an expansion of φ by further variables, i.e. instead of some constants $\bar{\varphi}$ unlike φ contains, in addition, only variables, i.e. to put it more exactly

$$\varphi = \mathrm{Sb}^{x_{n+1}, \ldots, x_m}_{t_{n+1}, \ldots, t_m}(\bar{\varphi})$$

and $\bar{\varphi}$ is, a maximum expansion of φ in the sense that it fulfils (*). Let us do this gradually for the whole Π_2 and let us define our new $\hat{F} : \Pi_1 \to \mathscr{P}(\Pi_2)$ in such a way that, instead of the above-mentioned sets of the shape $S(\varphi)$, which form the covering of the set $F(\pi)$, we shall supply the given formula $\bar{\varphi}$. The representation of $\hat{F}$, in fact, describes better the behaviour of subjects on the grounds that

1. $\hat{F}(\pi)$ contains fewer elements than
2. from each element of the set $\hat{F}(\pi)$ we can see what the subject disregards.

Moreover, it is possible from $\hat{F}(\pi)$ to pass easily to $F(\pi)$ by way od substitutions, and thus it can be said that there exist mapping ι_1, ι_2 such that diagram

$$\begin{array}{ccc} & F \nearrow & \mathscr{P}(\Pi_1) \\ \Pi_1 & & \iota_1 \downarrow \quad \uparrow \iota_2 \\ & \hat{F} \searrow & \mathscr{P}(\Pi_2) \end{array}$$

commutes (ι_1, ι_2 being, in fact, inverse functions).

Let us now examine again our example and let us form from F_1, F_2, F_3, F_4 the mappings $\hat{F}_1, \hat{F}_2, \hat{F}_3, \hat{F}_4$.

$$\hat{F}_1(\varepsilon_1 P_1(c_1, \gamma) \,\&\, \varepsilon_2 P_2(c_1, \gamma)) = \{\varepsilon_1 P_1(c_1, \gamma) \,\&\, \varepsilon_2 P_2(c_1, \gamma)\}.$$

(From a four-element set we have a one-element set.)

$$\hat{F}_2 = F_2, \quad \hat{F}_3 = F_3, \quad \hat{F}_4(\pi) = \hat{F}_3(\hat{F}_1(\pi)).$$

So far we have, as it were, studied only abstraction and generalization, as a single cognitive step in the process of identification. No less important a place, however, in identification procedures of our type (i.e. in the formation of meaning and sign) is understood as follows: in the foregoing inductive phase, making use of a hypothetical procedure and verifications, the subject formed a generalized cognitive scheme; this scheme has the character of a cognitive form, which necessarily implies that all those objects having in our research for example dots outside the ellipse belong to the sought, positively relevant class. This implication is connected with the origin of cognitive form (its origin and nature being still little accounted for) and is effected through operations on objects; it is a complex cognitive process, in which a figure newly presented, which the subject is to identify, is on the basis of its relevant quality placed in the class determined; at the same time an appropriate invariant meaning is assigned to the object in question (actually to its fundamental sign). This can be illustrated as follows: Let us assume by way of example that the subject found that π_1 and π_2 belong to the same class, and let Π_1 and Π_2 be of the following form:

$$\pi_1 = \pi_0 \,\&\, P(t_1, \ldots, t_{\mathrm{Ar}(P)}),$$
$$\pi_2 = \pi_0 \,\&\, \neg P(t_1, \ldots, t_{\mathrm{Ar}(P)}).$$

Then the subject can express a justified assumption that $P(t_1, \ldots, t_{\mathrm{Ar}(P)})$ is of no importance and accordingly, he has used the operation of immediate consequence defined in the following way

$$\frac{\pi \,\&\, \alpha, \pi \,\&\, \neg \alpha}{\pi}.$$

(This reads as follows: "From $\pi \,\&\, \alpha$, $\pi \,\&\, \neg\alpha$ one can derive π.") Besides the requirement of immediacy (see ALIOS [5]) another requirement is claimed here with respect to the operation of immediate consequence, namely, that the resulting elementary conjunction, which we shall obtain, should be shorter than the elementary conjunction from which we deduce. Thus we shall study *abstraction operations* of immediate *consequence*, which will be the subsets $\Pi_2 \times P_{\mathrm{fin}}(\Pi_2)$, where $P_{\mathrm{fin}}(\Pi_2)$ denotes a set of all finite subsets of the set Π_2, fulfilling the intuitive requirements of immediacy and reduction. Abstraction operations of immediate consequence shall be denoted AIC (if need be with an index) and shall be written φ AIC e or e/φ, where the finite set e is part of Π_2 and $\varphi \in \Pi_2$. Thus the above-mentioned example shall be written as a model in the following way:

$$\pi \text{ AIC } \{\pi \,\&\, \alpha, \pi \,\&\, \neg \alpha\}. \qquad (**)$$

Of what use is AIC? – AIC will greatly simplify the mapping of $\hat{F}$. Instead of $\hat{F}$ (which in itself is simpler than F) we shall study the mapping of F^*, which we shall obtain from $\hat{F}$ in the following way:

For each $\pi \in \Pi_2$ we shall form a system of subsets $e_1, \ldots, e_m \subseteq \hat{F}(\pi)$ such that $\varphi_1, \ldots, \varphi_m$ exist so that for $j = 1, \ldots$ we have φ_j AIC e_j.
Then

$$F^*(\pi) = (\hat{F}(\pi) - \bigcup_{j=1}^{m} e_j) \cup \{\varphi_1, \varphi_2, \ldots, \varphi_m\}.$$

Let us now examine, what will become of F_1^*, F_2^*, F_3^*, F_4^* formed from $\hat{F}_1, \hat{F}_2, \hat{F}_3, \hat{F}_4$ with the aid of AIC, which is described in (**).

$$\begin{aligned}
&F_1^* = \hat{F}_1,\\
&F_2^*(\varepsilon_1 P_1(c_1, \gamma) \,\&\, \varepsilon_2 P_2(c_1, \gamma)) = \{\varepsilon_2 P_2(c_1, \gamma)\},\\
&F_3^*(\varepsilon_1 P_1(c_1, \gamma) \,\&\, \varepsilon_2 P_2(c_1, \gamma)) = \{\varepsilon_1 P_1(c_1, \gamma)\},\\
&F_4^*(\pi) = F_3^*(F_1^*(\pi)).
\end{aligned}$$

Let us be aware of the fact that

1. for each π, $F^*(\pi)$ contains fewer elements than $\hat{F}(\pi)$ (and, accordingly, fewer than $F(\pi)$ as well);

2. for each π there are in $F^*(\pi)$ only those formulae from which all elements $\hat{F}(\pi)$ (and, accordingly, $F(\pi)$) can be "immediately derived". In $F^*(\pi)$ there will be (with suitable AIC) only those formulae which corespond to those psychological processes which make it possible for the solver to perform the act of generalization and the abstraction connected with it. This mechanism permits to effect not only the indispensable reduction of superfluous (redundant)

information, but also to create an *information invariant* corresponding to what is usually expressed by the term concept. In a problem situation set implicitly constants of meaning are substituted for the unknown X, which in our identification task is a synonym of the understanding of the principle of solving. In this concept, which comes into existence in the process of *learning* on the basis of the semiotic pair sign – meaning, relevant properties of cognized objects are brought together. These processes are a necessary condition for the subject to identify the objects so far unknown to him.

4. Summary

a) The study represents the first attempt made in this country at a simulation formalization of cognitive processes in *identifying* objects of figures of various complexity (though in our analysis we limited ourselves to the simplest possible figure, as seen in Fig. 1). Identification is not given beforehand, it originates in *learning* in a problem situation, which at the beginning is not clear.

b) A necessary condition for this simulation is the creation of a formal language which requires an exact definition of basic psychological terms concerning identification and classification. These terms should include: relevance of sign and meaning, concept, generalization abstraction and deduction. In the present study the concepts and cognitive processes in question are defined with the aid of *information analysis*. The basic psychological processes involved have been defined by us with respect to their different levels, which in a specific way participate in the identification procedure investigated.

c) Owing to a prospective *modelling* of those processes great importance is attached to the circumstance that the formal language proposed is based on extensive experimental data, which explain the nature of concept identification [12]. The model under discussion will be subject to verification on the basis of experimental data.

d) It is necessary to take into account that the approach described is one of the possible ones and that its justification can be assessed only on basis of further processing. As specified above, the mathematical-logical apparatus proposed by us is adequate with respect to our experimental results and with respect to the present state of psychology. The mathematical apparatus is simple and makes use of a basically very narrow restriction of the predicate calculus.

e) The study submitted represents only an introduction to the problems under investigation.

References

[1] W. R. Ashby: Kybernetika. Orbis, Praha 1961.
[2] J. S. Bruner et al.: A Study of Thinking. Wiley, N. York 1956.
[3] J. Deese, S. H. Hulse: The Psychology of Learning. Third edition. McGraw-Hill, N. York 1967.

[4] W. R. Gardner: Uncertainty and Structure as Psychological Concepts. Wiley, N. York 1962.
[5] P. Hájek: Automatic Listing of Important Observational Statements. Kybernetika *9* (1973), Part I 3, 187—205; Part II 4, 251—271.
[6] C. J. Hovland—W. Weiss: Transmission of Information Concerning Concepts through Positive and Negative Instances. J. exp. Psychol. *45* (1953), 175—182.
[7] A. Church: Introduction to Mathematical Logic. Princeton University Press, Princeton 1956.
[8] F. Klix: Information und Verhalten. Deutscher Verlag der Wissenschaften, Berlin 1971.
[9] V. I. Lenin: Materialismus a empiriokriticismus. SNPL, Praha 1952.
[10] J. Linhart: Psychologické problémy teorie učení NČSAV, Praha 1965.
[11] J. Linhart: Identifikace znaků a heuristické rozhodování. In: Kybernetické aspekty problémov identifikácie, Vysoké Tatry, 157—170.
[12] J. Linhart: Proces a struktura lidského učení. Academia, Praha 1972.
[13] J. Linhart: Hypoteticko-deduktivní modely a heuristické metody. Čs. psychologie *17* (1973).
[14] E. Mendelson: Introduction to Mathematical Logic. Van Nostrand Reinhold Co., N. York 1964.
[15] I. S. Narskij: Dialektický rozpor a logika poznání. Svoboda, Praha 1972.
[16] A. Newell, H. Simon: Human Problem Solving. Prentice Hall, New Yersey 1972.
[17] A. Quastler: Information theory in psychology. Illinois 1955.
[18] J. Piaget, B. Inhelder: Psychologie dítěte. SPN, Praha 1970.
[19] Д. А. Поспелов, В. Н. Пушкин: Мышление и автомати (D. A. Pospelov, V. N. Pushkin: Thinking and Automata). Советское радно, Москва 1972.
[20] В. Н. Пушкин: Психология и кибернетика (V. N. Pushkin: Psychology and Cybernetics). Педагогика, Москва 1971.
[21] S. L. Rubinštejn: O myslení a spôsoboch jeho výzkumu. SPN, Bratislava 1960.
[22] K. L. Smoke: Negative instances in concept learning. J. exp. Psychol. *16* (1933), 583—588.
[23] M. Stríženec: Psychológia a kybernetika. VSAV, Bratislava 1966.
[24] L. S. Vygotskij: Myšlení a řeč. SPN, Praha 1970.

GENERAL AND METHODOLOGICAL PROBLEMS OF ENTROPY AND INFORMATION

III

IS THE COARSE-GRAINED ENTROPY OF CLASSICAL STATISTICAL MECHANICS AN ANTHROPOMORPHISM?

Adolf Grünbaum

1. Introduction

To introduce the issue posed in the title of this essay, let me caution against a possible misconstrual of the classification "anthropomorphism" as a pejorative characterization of an attribute or entity postulated by a physical theory. It is clear enough that those who charge a *particular* concept employed in such a theory with being an anthropomorphism are *not* concerned to convey thereby the following truism: *Qua being a theory*, any known theory whatever is not only propounded and devised by man (or by a humanoid computer), but is also unavoidably *fallible* because of the postulational (inductive) risks inherent in the very logic of its intellectual construction. Thus, it is surely *not* the latter truism that E. T. Jaynes was interested in espousing when he wrote: "...entropy is an anthropomorphic concept, not only in the well-known statistical sense that it measures the extent of human ignorance as to the microstate. *Even at the purely phenomenological level, entropy is an anthropomorphic concept.* For it is a property, not of the physical system, but of the particular experiments you or I choose to perform on it."[1]) What may be overlooked, however, is that those who level the charge of anthropomorphism at a particular, specified ingredient of a physical theory ought *not* to be saddled with the following claim: For any and every item in such a theory, an unambiguous identification can always be made as to whether or not the item is an anthropomorphism within the postulational framework of the given theory. Nor should such a dichotomous claim be imputed to those who *deny* a *specific* charge of anthropomorphism in physics! This is not to say that the notion of anthropomorphism has not been employed in an avowedly dichotomous fashion, as for example in L. Kronecker's Pythagorean dictum "God made the integers; all the rest is the work of man."[2])

When considering the respective logical constitutions of the phenomenological and statistical entropies ascribed to a physical system, Jaynes singled out the extent of the role of human

[1]) [1, p. 398]; italics in the original. Jaynes mentions in his Acknowledgements that he first heard the remark "Entropy is an anthropomorphic concept" from E. P. Wigner.

[2]) Cf. [2, p. 170].

decision in the former and of human ignorance in the latter as touchstones for characterizing, or perhaps even indicting, them as anthropomorphisms. Whereas Jaynes was concerned to spell out his view of the bearing of human choice or decision on the status of the *phenomenological* entropy, I shall not deal with the latter kind of entropy at all. Instead, I shall focus on the significance of human volition in the logical constitution of the statistical coarse-grained entropy of classical statistical mechanics by addressing myself to the following question: Is the import of the role of human choice and decision for the *physical* credentials of that statistical entropy such as to sustain the charge of anthropomorphism against it?

As I shall set forth below in detail, such a charge might be made plausible by reference to the following fact: the occurrence and direction of a temporal change of the entropy $S = k \log W$ assigned to a given physical system depends essentially on *our human choice* of the *size* of the finite equal cells or boxes into which we partition the 6-dimensional position-velocity phase space ("μ-space"). In the pre-quantum classical statistical mechanics, there is enormous scope for the exercise of such choice in coarse-graining. For classical theory viewed action as a magnitude susceptible of having any value, and Planck's quantum of action h is unavailable in that theory to provide a *physical* basis for carving up the phase space into boxes each of which has the volume h^3.[3]) It is precisely because there is such relatively great latitude for human decision in coarse-graining the specified phase space of *pre*-quantum statistical mechanics that I wish to examine the soundness of the charge of anthropomorphism as leveled against the statistical entropy of that classical theory. It will then be a corollary of our analysis to determine whether the physical significance of that entropy can be validly impugned by regarding it as a measure of the extent of human ignorance as to the underlying microstate.

But before we can deal with the issue of anthropomorphism, we must give some detailed consideration to the aforementioned dependence of the temporal change of the entropy on the human decision as to the partitioning of the phase space into cells.

2. Entropy Change and Arbitrariness of the Partitioning of Phase Space

One of the things we must examine is the effect of *alternative* choices of a partition on the *direction* in which the entropy changes with time as the physical system develops. Let us briefly recall the meanings of some of the fundamental terminology amid explaining the special notation I shall use to indicate for each of the various pertinent quantities to what particular partition they are being referred.

Consider one particular system of particles of given finite total energy and given finite 3-dimensional volume. Call that system "α". Let the portion of 6-dimensional phase space whose occupation by the representative points of these particles is compatible with the given energy and physical dimensions be divided into a finite number m of equal cells $c_1, c_2, \ldots, c_m$.

[3]) Cf. [3, pp. 908—909].

Also let the number of particles which respectively occupy these various particular cells at a given time be $n_1, n_2, \ldots, n_m$. Such a set of numbers n_i $(i = 1, 2, \ldots, m)$ specifies the *macro-*state of the system by stating for each of the cells how many particles occupy it at the time. Thus the macro-state is characterized by the numerical *distribution* of the total finite number n of particles among the various particular cells, irrespective of which individual particles these may be. Suppose, for example, cells c_1 and c_2 were occupied respectively by unequal numbers of particles x and y at time t_1, but then were occupied by y particles and x particles respectively at time t_2. Then, since $x \neq y$, the macro-states (distributions) I and II at these two times t_1 and t_2 would be *different*, even though we assume that the numbers of particles in each of the remaining cells $c_3, c_4, \ldots, c_m$ had not changed at all. For the macro-state does depend on the identity of the *cells* among which differing numbers of particles are distributed, although the *macro*-state does not depend on the identity of the *particles* which are thus distributed.

The identity of the particles themselves is, of course, relevant to the *micro*-state (arrangement, complexion). The latter is specified by stating for each of the n particles in which one of the m equal 6-dimensional cells of the coarse-grained partition its representative point lies at a given time. By thus being specified only to within the finite cells of the chosen partition, the micro-state as such is *relative to the chosen partition*. By contrast, the *precise* micro-state of classical mechanics is given by the exact or punctal values of the six position-velocity coordinates and is therefore *independent* of the chosen partition. To distinguish the latter kind of micro-state from the former, I shall use the qualifying adjectives "precise" or "punctal." Although the unqualified term "microstate" will always refer to the partition-dependent, *non*-punctal kind of microstate, I shall occasionally emphasize that this relative kind of microstate is intended by using the adjective "non-punctal" or "coarse-grained."

For any one chosen partition, each kind of macrostate or distribution is clearly constituted by a certain class of microstates: Any occurrence of one of the members of this class of microstates constitutes an occurrence of the macrostate in question. The number W of different microstates belonging to the particular macrostate specified by the set of distribution numbers $n_1, n_2, \ldots, n_m$ is given by $W = \dfrac{n!}{n_1!\, n_2!, \ldots, n_m!}$. (It will be convenient in the sequel to denote the product of the factorials in the denominator of this expression by "$\pi n_i!$") And the entropy S assigned to any one macrostate to which W microstates belong is $S = k \log W$. Since increases or decreases of W are tantamount respectively to increases or decreases of S, we can conveniently work with W below rather than with S when calculating the directions of entropy changes. Indeed, it should be borne in mind that it will be permissible for our particular purposes to talk about W and S interchangeably.

Since multiplication in the product $\pi n_i!$ is commutative, our aforementioned two *different* macrostates I and II are each assigned the same entropy S. Indeed, *any* distribution of the n particles of the given system that gives rise to the same value of the product π will be assigned the same value of the entropy. Thus, one and the same entropy state S_0 of a given system can correspond to *different* distributions or macrostates *each* of which has W_0 microstates belonging to it, where the value of W_0 is given by $S_0 = k \log W_0$. Hence the number of different microstates which can underlie or belong to one and the same entropy state S_0 is given *not* by W_0 but rather by the *product* of W_0 with the number d_0 of different distributions having the same entropy S_0.[4])

Now let A and B be two different kinds of precise, punctal microstates in which the given physical system α may be at different times. Denote a particular partition of the corresponding phase space by the numeral 1. And let 1A and 1B be the respective coarse, *non*-punctal microstates of the system *with respect to partition* 1, when it is in the respective punctal microstates A and B. We shall generally be interested in cases in which the *non*-punctal microstates are different from one another in kind, no less than the punctal ones.

If we consider the characterization of the system in states A and B to within an *alternative* partition 2, then 2A and 2B are the respective non-punctal microstates of the system with respect to partition 2. Clearly, the one precise microstate A gives rise to the two different *non*-punctal microstates 1A and 2A, since the latter are specified only to within the cells of their different partitions. And similarly for B, 1B and 2B.

Returning to partition 1, we note that 1A and 1B each uniquely specify the distribution (macrostate) to which they belong with respect to partition 1. These distributions may or may not be the same, but in either case we shall denote them by "D1A" and D1B "respectively. Furthermore, we shall use "W_{D1A}" to represent the number of microstates belonging to D1A. For important reasons that we shall discuss later on, with respect to *the particular partition to which it pertains*, the quantity W_{D1A} does *triple duty* as follows: (i) As just noted, it is the number of different microstates belonging to the given distribution (macrostate) D1A, (ii) it is the (unnormalized) probability of occurrence of that distribution in the time ensemble of the macrostates which the system attains relatively to partition 1, i.e., the so-called thermodynamic probability of the macrostate in question, and (iii) it is a measure of the degree of *homogeneity* or evenness, equalization, disorder and "well shuffledness" of the given macrostate D1A in the class of macrostates which the system attains relatively to partition 1. This *triple significance* of the quantity W will turn out to have an important bearing on our central issue of anthropomorphism later on. Using the above notational devices, we shall say that S_{D1A} and S_{D1B} are the entropies of macrostates D1A and D1B respectively, i.e., the entropies of the system relative to partition 1 when it is in the respective precise microstates A and B. By the same token, the macrostates which A and B constitute relatively to partition 2 have entropies S_{D2A} and S_{D2B} respectively.

We shall now illustrate the following important sets of facts concerning the bearing of the chosen partition on the direction of the entropy changes of the given system α:

(i) There are partitions 1 and 2 of the corresponding phase space as well as precise microstates A and B such that if the system evolves from A to B,

$$S_{D1A} = S_{D1B} \quad but \quad S_{D2A} \neq S_{D2B},$$

[4]) The fact that $W_0 \,.\, d_0$ microstates rather than only W_0 of them underlie an entropy state S_0 was erroneously overlooked on p. 162 of my paper "The Anisotropy of Time" [4]. Hence the statement of boundary conditions governing the so-called branch systems given there must be amended: Either the random samples of the formulation must be particularized to each distribution of given entropy, as I shall do later on in this §2, *or* the random samples must be taken from the *total* number ($W_1 \,.\, d_1$) of microstates which can underlie an entropy state S_1 of the given kind of system.

i.e., although the system does *not* change its entropy relatively to partition 1 in the course of its transition from A to B, its entropy does change relatively to partition 2. Thus, the system's two macrostates D1A and D1B with respect to partition 1 are equiprobable, whereas its two macrostates D2A and D2B with respect to partition 2 have different probabilities. There is no inconsistency between these various probabilities.

(ii) There are partitions 3 and 4 as well as precise microstates E and F such that if the system evolves from E to F,

$$S_{\mathrm{D3E}} < S_{\mathrm{D3F}} \qquad \textit{but} \qquad S_{\mathrm{D4E}} > S_{\mathrm{D4F}},$$

i.e., although the entropy of the system *increases* relatively to partition 3 in the course of its transition from E to F, it *decreases* relatively to partition 4. Since these two entropy changes have opposite signs, the *probability ranking* of D3E and D3F will be opposite to that of D4E and D4F. There is no inconsistency between these various probabilities.

To make our diagramming and arithmetic quite simple, we shall represent the 6-dimensional phase space of the system α by a mere rectangle, and we shall choose four convenient numbers m of equal boxes (cells) to generate the four different partitions of the representative rectangle. Also, it will simplify our formulation without detriment to rigor to speak of the particles themselves rather than of their representative points as occupying certain cells in phase space.

Case (i)

We shall, of course, not label and depict the individual n particles, since we shall not diagram the precise microstates A and B but only their corresponding respective pairs of macrostates relative to partitions 1 and 2. The first partition will consist of two boxes ($m = 2$), while the number m of boxes of partition 2 will be 4.

A macrostate relative to a given partition is specified *only to within the particular cells* of that partition. Hence the shading of a given box in the diagram will signify that the positive integral number of particles appropriate to the given distribution of particles are to be found *somewhere or other* in that box. If a given box is not shaded, then it contains no particles *anywhere* within it.

Let the precise microstate A be such that the first of the two boxes of partition 1 contains half of the particles *somewhere within it*, while the second, of course, contains the remaining half of the n particles *somewhere within it*. Thus D1A is specified by the sequence $n/2$, $n/2$. Let B be a precise microstate different from A but such that the macrostate D1B which it generates relatively to partition 1 is the same as D1A. Thus, we diagram:

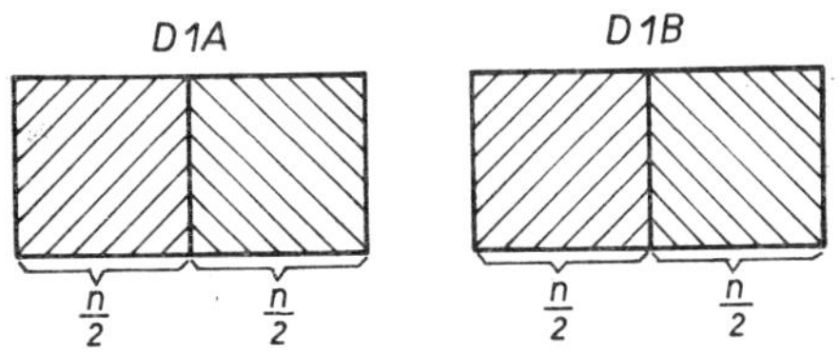

Since A and B generate identical distributions D1A and D1B relatively to partition 1, their thermodynamic probabilities relative to that partition will have the same value, viz.,

$n!/((n/2)!)^2$. Hence relatively to partition 1 the system will have the same entropy when it is in punctal microstate A as when it is in the different punctal microstate B, and we can write

$$S_{D1A} = S_{D1B},$$

as claimed initially.

Turning to partition 2 ($m = 4$), it is clear that A and B and the integral number n of particles can also be such as to generate the following quite different distributions relatively to partition 2: In macrostate D2A, boxes 1, 2, 3 and 4 respectively contain $n/2$, 0, $n/2$ and 0 particles, whereas D2B involves the uniform presence of $n/4$ particles in each of the four boxes:

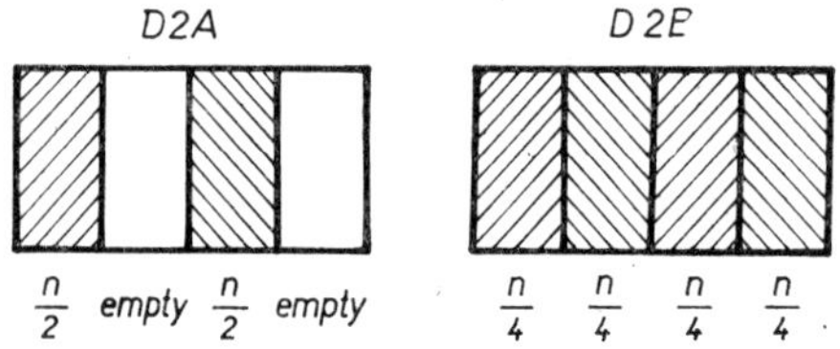

A and B can and must be so chosen that the requirement of energy conservation and other relevant constraints on the closed system α are indeed satisfied.

Hence we have

$$W_{D2A} = \frac{n!}{\frac{n}{2}!\,0!\,\frac{n}{2}!\,0!} = \frac{n!}{\left(\frac{n}{2}!\right)^2},$$

a value which, incidentally, is the same as that of W_{D1A} and W_{D1B} above. But

$$W_{D2B} = \frac{n!}{\left(\frac{n}{4}!\right)^4}.$$

Since $((n/4)!)^4 < ((n/2)!)^2$, the quantities W pertaining to the different macrostates D2A and D2B satisfy the inequality $W_{D2A} < W_{D2B}$, so that we can write

$$S_{D2A} \neq S_{D2B},$$

as claimed initially.

We have illustrated the joint validity of the relations

$$S_{D1A} = S_{D1B} \qquad \text{and} \qquad S_{D2A} \neq S_{D2B}$$

by reference to a case in which $m = 2$ for partition 1, while $m = 4$ for partition 2. Thus, in our particular illustration the entropy of the system is *constant* for the transition from A to B relatively to the partition possessing the *smaller* number of cells. Thus, our particular illustration is one in which the entropy *changes* relatively to the partition having the *larger* number of cells. But it is easy to provide another illustration of the joint validity of the above entropy relations in which the *constancy* of the entropy obtains relatively to that partition which has the *larger* number of cells. Hence consider the *general* case in which a transition of the system from A to B involves no entropy change relatively to one of two partitions while producing an entropy change relatively to the other. Then we can say that the *change* in entropy is *not* generally correlated with the partition containing the larger number of cells. In other words, the success or failure of one of two partitions in effecting an entropic differentiation between A and B is *not* generally a matter of the relative coarseness of the partition.

It is patent that there is no inconsistency between (1) the assertion $W_{D1A} = W_{D1B}$, when construed as claiming the *equiprobability* of the two macrostates D1A and D1B, and (2) the assertion $W_{D2A} \neq W_{D2B}$, when interpreted as affirming that the distributions D2A and D2B have *different* probabilities of occurrence in time. There are two different time ensembles of macrostates which our physical system α exhibits relatively to the two different partitions 1 and 2. In one of these two time ensembles, the two *different* distributions D2A and D2B respectively generated by A and B relatively to partition 2 can readily have *unequal* probabilities of occurrence, while the identical distributions D1A and D1B which A and B generate relatively to partition 1 will, of course, have one and the same probability. For the different probability *rankings*, i.e., the respective unequal and equal probabilities corresponding to these two partitions, are *not* ascribed to the pair of precise microstates A and B themselves but only to the two different pairs of macrostates D1A, D1B and D2A, D2B which are associated with the one pair A and B relatively to the two differing partitions. The consistency of the two different probability rankings corresponding to the two partitions we have diagrammed above is made palpable by these diagrams as follows: It would be a blatant physical falsehood to claim that the different macrostates D2A and D2B depicted in our second diagram are equiprobable, but it is plain that the two identical distributions of our first diagram are equiprobable.

A further important facet of the bearing of the chosen partition on the entropy changes of the system α will now emerge from illustrating the *oppositely directed* entropy changes

$$S_{D3E} < S_{D3F} \qquad \text{and} \qquad S_{D4E} > S_{D4F}$$

stated as (ii) above.

Case (ii)

I am indebted to Allen Janis for the particular illustration which is about to follow and for a very helpful discussion of its significance for the entropy statistics of ensembles of systems.

Let partition 3 consist of 9 boxes, and let the precise microstates E and F as well as the number n of particles be such that the following two distributions are generated in the course of the system's transition from E to F: D3E is constituted by the presence of *all* of the n particles in the first box of partition 3, while D3F is a uniform distribution of the particles among the 9 boxes, so that each of the 9 boxes of partition 3 contains $n/9$ particles. Hence we have

$$W_{D3E} = \frac{n!}{n!} = 1 \quad \text{and} \quad W_{D3F} = \frac{n!}{\left(\frac{n}{9}!\right)^9}.$$

Without computing the value of the second of these two probabilities, it is evident that many more than 1 non-punctal microstate belong to D3F, while only 1 such microstate belongs to D3E. Therefore, $W_{D3E} < W_{D3F}$, and $S_{D3E} < S_{D3F}$, as claimed initially.

Without specifying the actual number m of cells in partition 4, let that number (as well as the states E and F) be such that D4E and D4F have the following respective properties: In macrostate D4E, each box of the comparatively fine partition 4 contains *at most one* particle, whereas in macrostate D4F, each cell of this partition contains either two particles or none. Note that in the case of D4F, the n particles are so distributed that there are $n/2$ cells each of which contains exactly two particles while the remaining cells are each empty. Hence we have

$$W_{D4E} = \frac{n!}{1} = n! \quad \text{and} \quad W_{D4F} = \frac{n!}{(2!)^{n/2}}.$$

Evidently $W_{D4E} > W_{D4F}$, so that $S_{D4E} > S_{D4F}$, as claimed initially.

Therefore, during the transition of the system α from E to F, the entropy relative to partition 3 *in*creases whereas the entropy with respect to partition 4 *de*creases. For reasons of the kind discussed under Case (i) above, the probability ranking $W_{D3E} < W_{D3F}$ is entirely *consistent* with the *opposite* probability ranking $W_{D4E} > W_{D4F}$.

Joseph Camp has given me an illustration of the existence of partitions 5 and 6, and of a temporal sequence of precise microstates A, B, and C such that both

$$W_{D5A} < W_{D5B} < W_{D5C} \qquad \text{and} \qquad W_{D6B} < W_{D6C} < W_{D6A}.$$

We have set forth the *relativity to the chosen partition* of the very occurrence *and* direction of an entropy change in *one* particular physical system α during its transition from one punctal microstate to another. This relativity to the chosen partition for a single system α raises the important question whether the *statistics* of temporal entropy change in large but finite *ensembles* of like closed systems are similarly partition-dependent and might thereby be claimed to depend indirectly on human choices.

To deal with this question, let us be more precise as to the nature of the ensembles of systems to which it pertains. The systems in each and every ensemble will be alike: Each system will consist of, say, the same kind of gas confined to regions of three dimensional space having the same volume, and possessing the same total energy. In *one* of these ensembles, say K_1, to which our one system α above belongs, each system is in an initial state which is the same kind of distribution (macrostate) as D3E relatively to partition 3, and the same kind of distribution as D4E relatively to partition 4. But the precise microstates which underlie these same initial distributions in the various members of K_1 are emphatically *different* in kind from E in all except α, i.e., the precise microstates all differ from one another in kind. Moreover, the respective non-punctal microstates which generate the initial macrostate D4E in the various systems of K_1 will *each* be a *random sample* of the FINITE set of W_{D4E} different microstates compatible with D4E in the pertinent member of K_1. I recognize that the concept of a random sample or of a random selection is not wholly unproblematic as characterized in the following standard kind of definition: "A sample obtained by a selection of items from the population is a *random sample* if each item in the population has an equal chance of being drawn. *Random* describes a method of drawing a sample, rather than some resulting property of the sample discoverable after the observance of the sample." [5, p. 300] But I assume that such difficulties as may beset the concept of a random sample in its application to *finite* sets—which is the application I have made of it here – can be resolved as follows: The clarified concept of random sample accords with the physical correctness of the statistical claim that I shall make below concerning the entropic behavior of the ensembles of systems which we are engaged in discussing. In the case of the initial distribution D3E, there is only one underlying microstate for each system since $W_{D3E} = 1$, and hence this random sample requirement is trivially satisfied. All the members of K_1 have, of course, the same initial entropy S_{D3E} relatively to partition 3, and also the same initial entropy S_{D4E} with respect to partition 4.

A *second* ensemble, K_2, resembles K_1 in that the initial macrostate of its members relative to partition 3 has the same entropy S_{D3E}, and similarly for the initial macrostate of entropy S_{D4E} relative to partition 4. But the members of K_2 differ from those of K_1 in that their initial macrostates relative to our two partitions are constituted by distributions which differ in kind from D3E and D4E. Except for the latter difference in the characteristics of K_1 and K_2, the other statements I made about the initial conditions governing the members of K_1 apply, *mutatis mutandis*, to those of K_2. By the same token, we can consider further ensembles of systems $K_3, K_4, \ldots, K_n$, where n is a finite integer, in *each* of which the systems start out in entropy state S_{D3E} and in entropy state S_{D4E} but with the following difference between the various ensembles: Relative to a *given* partition, the kind of distribution which constitutes the initial entropy state of the systems in *one* of the ensembles *differs* from the kind of distribution constituting the self-same initial entropy state of the systems in any *other* ensemble. Furthermore, the cardinality of each ensemble is finite though very large, say $n!$, where n is the number of particles in each system and is of the order of 6×10^{23}. Therefore the union U of our ensembles will contain only finitely many systems. And hence it will be meaningful to make an assertion about a *majority* of the members of U.

We have been characterizing our ensembles with respect to two particular partitions, 3 and 4. But clearly, we can introduce at least a denumerable infinity of other partitions and can characterize our ensembles relatively to them, *mutatis mutandis*, in the same terms as we did with respect to 3 and 4. Hereafter we shall therefore not, in general, be restricted to any particular finite number of partitions.

The precise microstates E and F of our primary single system α were separated by some particular time interval Δt. We saw that during α's transition from E to F in the time Δt, the direction of its entropy changes was partition-dependent: the entropy of α increased relatively to partition 3 while decreasing with respect to partition 4. Let us now consider each of the at least temporarily closed member-systems of our various ensembles $K_1, K_2, \ldots, K_n$ at two different times t and $t + \Delta t$ in order to point out a fact of basic importance to our concerns: There is complete compatibility between the partition-*dependence* of the direction of the entropy changes undergone by α, on the one hand, and, on the other hand, the following partition-*invariance* of the temporal statistics of entropy change in the union U of our ensembles of systems: For any and every partition, the common initial entropy of all the systems in U will either have increased *in a vast majority* of the systems by the time $t + \Delta t$, or it will have remained the same.[5]) Note that we here used the indefinite article "a" rather than the definite article "the" when speaking of "*a* vast majority of systems".

For it is crucial to emphasize that the partition-*invariance* of an entropy increase or of entropy constancy is being asserted only in the sense that for every partition, *some majority or other* of the systems in U will exhibit the specified entropic behavior. Thus, the partition-invariance of the stated *entropic behavior* of a majority M of the systems in U does *not* at all require that the *membership* of the set constituting M relatively to a particular partition be itself partition-invariant! Evidently, the stated partition-invariance of the statistics of entropy change fully allows our finding above that during its transition from E to F, the single system α underwent specified oppositely directed entropy changes relatively to partitions 3 and 4. The latter finding does illustrate that the *membership* of M is NOT partition-invariant as follows: Relatively to partition 3, the particular system α does belong to M, whereas relatively to partition 4, α belongs to the *minority* of systems which decrease their entropy during Δt.

The claim that α thus belongs to what is only a *minority* of systems undergoing an entropy decrease relatively to partition 4 appears to be quite justified. The reason is that even within the single ensemble K_1, which is included in U, in all likelihood *most* systems that started out in macrostate D4E with entropy S_{D4E}, will *not* have evolved, after an interval Δt, into any distribution which is such that each cell of partition 4 contains either exactly two particles or zero particles; nor into any other distribution possessing the lower entropy S_{D4F} of the latter kind of distribution; nor yet into any distribution whose entropy is lower than S_{D4E}. Relatively to partition 4, only a certain minority of systems decrease their initial entropy S_{D4E} during Δt, and α does belong to that minority with respect to this particular partition. But relatively to partition 3, a *different* minority of systems – though still only a minority – decrease their initial entropy S_{D3E} during Δt, and α does *not* belong to that minority relatively to the latter partition; instead, with respect to partition 3, α increases its entropy. Thus, our partition-dependent findings concerning the system α accord entirely with the partition-invariance of the entropy statistics for the class U of systems. *Mutatis mutandis,* the same partition-invariance holds for the finitist entropy statistics in the respective unions of ensembles of *other* kinds of *like* closed systems.

Hence we draw the following very important conclusion: Although the direction of the entropy change exhibited by a single system such as α is indeed relative to the chosen partition, *no such relativity to the chosen partition characterizes the statistics of entropy change in the specified ensembles of systems.* For there is an invariance in the entropic behavior of a majority of the systems in U amid the relativity to the partition of the membership of that majority. I have called attention to the partition-invariance of the latter statistics of entropy change in order to discuss the bearing of the *physical significance* of this invariance on our central issue of anthropomorphism. Before we can turn to giving a statement of that bearing, we must deal more explicitly than above with the following triple status of the entropy: As

[5]) For a statement of the physical reasons underlying this statistical claim as such, see pp. 162—163 of my "The Anisotropy of Time" [4].

we recall, it is the logarithm of a number W which is a measure of the degree of homogeneity as well as of the probability of occurrence, and not just the number of underlying microstates.

3. What is the Physical Significance of the Triple Role of the Entropy for the Entropy Statistics in the Class U?

For the time being, our attention will be devoted to an individual physical system and only later will we turn to our class U of systems.

(i) Entropy as a Measure of the Degree of Homogeneity

Perhaps it is possible to develop a notion of the degree of homogeneity or equalization of a distribution which is *not* relativized to a partition of the phase space. At least in regard to the speeds of the particles, the possibility of such a measure is suggested by the fact that the Maxwell distribution of molecular velocities for the *equilibrium* state of uniform temperature gives a *partition-independent maximum spread* among the speeds of the particles, the spread being maximum within the confines of the fixed total energy of the system. And perhaps some function depending on the spatial distances or spatial spread among the n particles as well as on their velocity spread could be introduced to provide a partition-independent measure of homogeneity for the various macrostates of the system. To my knowledge, no such partition-*independent* measure of homogeneity has been worked out for 6-dimensional position-velocity space. In this context, the terms "homogeneity" and "disorder" are synonymous.

In any case, let us employ a concept of degree of homogeneity which is relativized to a partition. Hence we inquire whether the intuitively suggested, avowedly partition-dependent attribute of degree of homogeneity of a distribution D is rendered numerically by the quantity $W = n!/\pi n_i!$, *initially* construed as merely the number of different microstates belonging to D. After mentioning some obvious reasons for an affirmative answer to this question, I shall call attention to cases in which our intuition of *qualitative* relative homogeneity gives no clear cut verdicts, in order to point out that I see no conflict between intuition and the homogeneity rankings furnished by the function W for these cases.

As for the obvious reasons, we observe that W is indeed maximum when the n particles are evenly allocated to the various cells, as compared to more or less *un*even allocations among the cells of the given partition. Furthermore, two distributions, which differ only in that the sequence of particular, generally unequal numbers $n_1, n_2, \ldots, n_m$ which specifies one of them is obtained by some permutation of the particular numbers specifying the other, clearly have the same degree of homogeneity on intuitive grounds. For intuitively, the degree of homogeneity does *not* depend on the *identity* of the particular cells in which given numbers of particles are distributed. And hence by assigning the same number to two distributions which differ in only this way, the function W implements our intuitive idea of relative degree of homogeneity in such cases.

But we can see that the function W also assigns the same number to certain distributions which differ far more strongly from one another. To see this, note first that for a sufficiently large number of cells ($m \geqq 4$), any given value of the product $\pi n_i!$ ($i = 1, 2, \ldots, m$) resulting from one particular distribution is capable of various decompositions into factorials, such that $\sum_{i=1}^{i=m} n_i = n$ *and* such that the sequences of factorials in the different decompositions are *not* obtainable from one another by permutation. An arithmetically simple example of this kind of nonunique decomposition into *different* sets of m factorials is furnished by the physically uninteresting case of $n = 7$. For note that $24 = 4!1!1!1! = 3!2!2!0!$, while the condition $\sum_1^m n_i = n$ is satisfied, since $4+1+1+1 = 3+2+2+0 = 7$. Thus, the function W assigns the same degree of homogeneity even to some distributions which differ such that they are *not* obtainable from one another by moving the various *fixed* numbers of particles into other cells of the partition. Yet it seems to me that our qualitative intuition of degree of homogeneity does not interdict the verdicts of equihomogeneity furnished by W for distributions that differ from one another as just specified.

Thus, we can say that the numerical measures furnished by the function W render our *qualitative* intuitive idea that the relative degree of homogeneity of a distribution with respect to a given partition P is determined by *how evenly* the particles are distributed among the cells of P, the degree of evenness being a matter of how many different cells contain more or less nearly equal numbers of particles. And since we can speak of W and S interchangeably for our purposes, we can say that the entropy S provides a measure of the degree of homogeneity.

(ii) Entropy as a Measure of the Probability of Occurrence of a Distribution

Our concept of degree of homogeneity is relativized to a partition, and that degree is indeed measured by the entropy. Hence to say that a macrostate is a state of high entropy does tell us that it is a highly homogeneous distribution relatively to the partition with respect to which the given macrostate is both defined and of high entropy. But the assertion that high entropy states are states of great homogeneity does *not* thereby become a tautology. For to say that a macrostate is a distribution D of high entropy relatively to a partition P is to say much *more*, physically speaking, than that D is highly homogeneous relatively to P: It is to say *as well* that D has a high probability of occurrence in the time ensemble of macrostates which the given system attains relatively to the partition P. Hence the entropy is a property of D which has at least the following objective macroscopic physical significance: It links or correlates the probability of the occurrence of a macrostate in time with its degree of homogeneity. I do not see how the physical significance of this important connection can be gainsaid, just because the entropy can *also* be regarded as a measure of the extent of human ignorance as to the actual microstate in the following sense: In any given occurrence of a certain macrostate D, only *one* of the W microstates that belong to D is actually being physically realized, but we don't know which one it is.

Hence I deny that W, as ascribed to the macrostate of the system at a given time, has a significance which is confined to the fact that, for all we know, *any one* of W different microstates is actually being realized at the time.

Even in its role as the mere number of different microstates belonging to a given kind of distribution D, the quantity W plays a physically relevant role and seems to me to be misleadingly belittled by being dubbed a measure of human ignorance. My reason for this claim will become clear after first noting the following: As is vouchsafed by the Birkhoff-von Neumann quasi-ergodic hypothesis, each one of the m^n microstates of which the permanently closed system is capable actually occurs *with the same frequency in time* or has the same probability $1/m^n$. And this ascription of equiprobability to all microstates is therefore *not* an assertion expressive of our lack of knowledge as to the actual (relative) frequency of the different microstates. In other words, having been justified by the quasi-ergodic hypothesis, this claim of equiprobability of microstates does *not* rest on an argument from *in*sufficient reason, akin to the Laplacian's invocation of the principle of equiignorance (indifference) in defense of his *a priori* probability metric for dice. Let me therefore take *this* physical and epistemological status of the equiprobability of all microstates for granted. Then I can justify my contention above that even qua just being the number of different microstates belonging to a given distribution D, the quantity W is physically relevant, although only one of the W microstates belonging to D is being realized in any given occurrence of D. For note that since all microstates of which the closed system is capable actually occur equally often, the number W of microstates belonging to a given kind of distribution D *makes for* and determines the actual frequency of that macrostate D in the time ensemble of the system's macrostates.

So much for the physical and epistemological credentials of the entropy as an attribute of the macrostates of an individual physical system. Let us now turn to our issue of anthropomorphism by using the results of our analysis to assess the *physical significance* of the statistics of entropy change in our class U. As will be recalled, the focus of our interest is whether the role of partition-dependence and thereby of human choice of the partition is such as to vitiate the physical significance of these statistics of entropy change and to relegate these statistics to being expressive of an anthropomorphism.

4. Do the Roles of Human Decision and Ignorance Impugn the Physical Significance of the Entropy Statistics for the Class U?

We noted the fundamental presumed fact that in the class U of systems, there is the following kind of partition-*invariance* of entropy increase or entropy constancy: For any and every partition, the common initial entropy of all the systems in U will either have increased in a vast majority M of the systems by the time $t + \Delta t$, or it will have remained the same. The substitution of one partition for another affects *only the membership of M* while leaving the specified partition-invariance of the entropy statistics intact. On this basis, we can now proceed to characterize several states of affairs as objective macroscopic physical facts, i.e., as *neither generated by a human choice of a partition nor expressive of human ignorance.* It is to be understood, of course, that the impending ascriptions of factual physical status are relative to the metrics of space and time ingredient in any of the partitionings of the 6-dimensional phase space and in the total volume and energy of each system.

It is a presumed fact that for any and every partition P, most systems in U which start out in a macrostate that is *comparatively inhomogeneous* relatively to P, after the time interval Δt will be in macrostates which are more homogeneous with respect to P as measured by W or S. The relativity of homogeneity itself to a partition P does not make it any less an objective physical fact that, by the time $t + \Delta t$, most members of the specified subclass of U will be in macrostates which *are* more homogeneous relatively to P. Moreover, these more or less homogeneous macrostates have the property of occurring more or less frequently in the respective time ensembles of macrostates of their respective systems. And the fact that the lower and higher frequencies (probabilities) of occurrence pertain to distributions which are specified relatively to a partition does not detract from the objectivity of these frequencies. Nor are these various physical facts demoted to the status of anthropomorphisms just because they pertain to macrostates, and, as such, do not comprise the *still richer* factual physical content envisioned by *punctal* mechanics or fine graining.

Thus, the statistics of entropy change in U codify partition-invariant macroscopic physical facts and cannot be held to be merely expressive of human ignorance of the underlying microprocesses. Human scientists do single out one or another particular partition relatively to which they characterize the macroscopic behavior of one or more systems entropically. But the partition-dependence of the direction of an entropy change in any one system no more makes the change of macrostate in that system anthropomorphic than the dependence on the inertial frame of the spatial distance between two events in Minkowski space-time renders that spatial distance anthropomorphic.

The gravamen of the charge of anthropomorphism which I have considered was that such partition-dependence as does obtain in the entropic descriptions of classical statistical mechanics renders these descriptions anthropomorphic. I have tried to argue that this charge is unfounded and that, at least to this extent, the use of the entropy concept is not illustrative of what J. L. Synge has aptly called the "Pygmalion syndrome" [6, p. 8].

Acknowledgments

It is a pleasure to acknowledge the benefit of discussions with Allen Janis, Robert B. Griffiths, Joseph Camp and Richard Creath, as well as the support of research by the National Science Foundation (USA).

References

[1] E. T. Jaynes: Gibbs vs. Boltzmann Entropies. American Journal of Physics *33* (1965), 5, 391—398.

[2] E. T. Bell: The Development of Mathematics. 2nd edition. McGraw Hill Book Co., New York 1945.

[3] A. d'Abro: The Rise of the New Physics. Dover, New York 1951.

[4] A. Grünbaum: The Anisotropy of Time. In: The Nature of Time (T. Gold, Ed.). Cornell University Press, Ithaca, N. Y. 1966, 146—186.

[5] R. C. James and G. James: Mathematics Dictionary. 3rd edition. D. Van Nostrand Co., Princeton, N. J. 1968.

[6] J. L. Synge: Talking About Relativity. North Holland Publishing Co., Amsterdam and London 1970.

THE PROBLEM OF THE OBJECTIVITY OF INFORMATION

A. D. Ursul

It has become obvious that the question under discussion in this article is important for philosophy as well as for other scientific branches which use the notion of information and the mathematical methods applied for its study. Philosophers are seeking the categories in which this new scientific notion could be comprehended and are also studying its relations to the notions of existence and spirit, matter and consciousness, i.e. to the basic notions of every philosophical trend. The solution is by far not without interest to scientists in the field of cybernetics, in the theory of information, and in the applied scientific fields because it underlies the methodology of scientific research in all these fields. Therefore, the question of the objectivity of information becomes the key question in the elaboration of the philosophical aspects, of the theory of information.

1. Fundamental Question of Philosophy and the Problem of Information

Information is discussed by journalists as well as by linguists, specialists in the field of cybernetics, psychologists, biologists, physicists, chemists, astronomers and scientists working in many other fields of the sciences. Basically, the problem of information has become one related to all branches of the sciences in general and the theoretical method of information has exceeded the framework of the theory of communication and cybernetics. The term "information" has become extremely fashionable and authors are wont to apply it in a broad sense and in the most varied contexts.

This wave of interest in the notion of information is easily understandable when we consider that we are, in fact, living in a century of the scientific and technological revolution which bears significant effect on all other spheres of human activity. One of the fundamental and most important aspects of the scientific and technological revolution of the second half of the 20th century is the very interest of science in the study of information and the interest of all technical fields in building up informational control (cybernetic) systems. Power production, the processing of various materials and the modern production process in general cannot exist without the movement of information.

With scientific and technical progress the notion of information becomes one of the leading problems and also one of the most topical problems of the day. It is evident that philosophy centres its attention, to varying degrees, on the most important problems of the individual scientific branches. The process and state of the study of the problem of information serve to prove that philosophy is by no means indifferent to the requirements of the other scientific branches.

At the same time it is well known that the problems and results of the individual branches of the sciences, mainly natural sciences, can always be interpreted from various philosophical standpoints. The relation between the results attained by individual scientific branches and their philosophical interpretations is, however, not unambiguous. The same is true of the relation between the notion of information and the other notions of cybernetics – they have aroused the interest not only of the representatives of Marxist-Leninist philosophy but also of scientists representing the other philosophical trends. It is obvious that different and in many cases contradictory trends strive to prove that various notions, such as that of "information" itself, correspond to their system of philosophical categories and concepts and to no other. Currently, not even the most abstract idealistic systems can ignore this notion and strive to find in it a basis for their respective system.

However, the notion of "information" itself "reacts" in a very diverse manner to the endeavours of diverse contradictory philosophical trends to interpret it and to a certain extent to assimilate the ideas linked with it. The theory of information and cybernetics are currently undergoing such stormy progress that we shall not have to wait long for the confirmation or refutation of philosophical concepts. Ten years of the development of modern sciences will suffice for the absorption of fruitful philosophical ideas and for the rejection and merciless elimination of the useless ones.

As for the notion of information, various and in many cases incompatible philosophical concepts were introduced shortly after the notion itself. The number and scope of these concepts most probably exceeds anything that had ever been done in this sphere. The scope of this article does not allow for a broad discussion on even the most general of these concepts and we have to restrict ourselves to those concepts which are directly related to our subject.

The problems of the objectivity of information is one of the relation of philosophical trends to this notion, a problem that is closely linked with the fundamental problem of philosophy. Different views have been expressed on the concept of information, corresponding to the respective philosophical systems. Thus, for instance, the representatives of the idealistic trend in philosophy mystify information when they present it either as a virtually subjective phenomenon or when they recognize its objective character only as one that is a manifestation of some spiritual origin.

The representatives of the latter trend, mainly the adherents of Thomas Aquinas wrote many theses on information. They emphasize that the conclusions they have arrived at do not differ from data acquired by present-day science, except for their assertion that all that exists, including information is the work of God. Information is being assigned a special role amidst all notions used by the sciences – namely, that of the intermediary between God and the material world.

These are scholastic and mystical constructions as the idea of spiritual reality has not been confirmed by the process of scientific cognition. The idea of God has not been confirmed by either information or cybernetics, both of which are founded on the fact that information is a property of matter and in this sense it is objective. Another, fantastic meaning of the notion of the objectivity of information, is assigned to it by the representatives of the neothomistic trend in philosophy and has proved to be unacceptable for current sciences which use the notion of information and apply its methods of study. The assumption of the transcendental character of information is not applied by specialists in cybernetics or in the theory of information in any manner. Not even in the works of those cybernetic specialists who are adherents of neothomistic views on the intelligible substance of information are these views proved in any way and they do not by any means serve as methodologically effective postulates in concrete scientific research.

The theory of the transcendental character of information, a process that takes place outside the human senses, has not brought any benefit to present-day science and is practically ignored by natural scientists and cybernetic experts. This concept thus has only a religious and philosophical impact and has been applied in the most recent experiments in the galvanization of theological and ecclesiastical dogma. The history of science thus enables us to pronounce a just sentence on one philosophical concept of the basis of information. The trend of the genuine scientific comprehension of the substance of information, as we shall further show, lies outside the objectively-idealistic philosophical systems.

A similar conclusion may be arrived at in another idealistic trend in philosophy – the subjectively idealistic trend. It is a well-known fact that a number of representatives of subjective idealism, especially the existentialists and the neopositivists have expressed their views on the problem of information.

The existentialists are lured to the notion of information by its alliance with the notion of communication which is extremely important in their philosophical system. Because they focus their attention on the mystified notions of human existence and communication between individuals (which is comprehended in a subjectively idealistic way) the existentialists conceive the notion of information in the pre-cybernetic spirit, namely as a category of communication, of contact between human individuals, in the spirit of extreme individualism. The existentialists conceive the world in general as a cryptogram, coded in a language that is incomprehensible to man, in short, man finds himself in a world of cryptic symbols and signs. The existentialist variant of the "symbol concept of man" in fact follows upon the subjectivist-idealistic concept and the use of the cybernetic notion of information is only superficial. In this case the very word information (as a fashion) is more important than the ideas of current scientific research that are connected with it. In fact, existentialism only develops the oldest concept of information and what more it enlarges excessively the subjectivism and idealism of this original concept. It no more than plays around with science, juggling with words – the existentialist concept has no relation whatsoever to cybernetics and to the information theory. The concept of the subjective character of information in the philosophy of existence has long ago been overcome by cybernetics and does therefore not affect the latter's progress.

Another variant of the subjectively idealistic approach to the substance of information has been presented by the adherents of positivism. It should be noted that a number of the representatives of this trend have contributed to the development of the theory of information. Here we should mention the development of the semantic theory of information in the works of R. Carnap, Y. Bar-Hillel and some of their successors. The works of these authors have considerably contributed to the further elaboration of the semantic concept of information which helped to eliminate the shortcomings of the original theory. That theory had applied a very poor formalized language, containing only the final number of names of individuals and independent one-figure predicates. Apart from this, in accordance with the concepts of logical semantics they only formalized the meaning of the object (i.e. the relation of the symbol to the objects which it described), while abstracting it from the sensual meaning. Last but not least, this theory only studied the hypotheses and other probability forms of cognitive processes, applying the means of inductive logic, while concrete scientific information as a rule has the character of a verified fact.

All these shortcomings of the original variant of the Carnap and Bar-Hillel theories are being overcome in the works of Kemeny, Wells, Maloney, Voyshvillo, Shreyder and of other scholars. Yet even today there does not exist a semantic concept of information which could be applied in the analysis of the contents of human information in such a manner as the Shannon theory is applied in the analysis of the formal (syntactic) properties of the information.

In itself the study of the semantic properties of information is certainly a very fruitful and promising trend in the information theory which is, however, still in its initial stage. The trend does not exceed original concepts of information in that it studies information transmitted by human speech (simplified for purposes of formalization).

The development of the semantic theories of information the same as the development of pragmatic concepts contributes little or nothing to the progress of concepts of the substance of information, especially from the point of view of its objectivity. We have to exceed the framework of semantic and pragmatic concepts or to extend these concepts in such a manner as to make them usable even for the description of biological informational processes (there are very few works dealing with this question). It is for this reason that the solution of the problem of the objective or subjective status of information is based on the interpretation of mathematical concepts of information and the application thereof.

What are the interpretations offered us by neo-positivism? Much depends on the respective trend to which this or that philosopher adhers, whether it is the logical or the linguistic trend. They all have one aspect in common, namely, that information is conceived subjectively or intersubjectively. Information is presented as an aspect, most frequently as the content of information conveyed by a symbol which in turn represent a certain characteristic or the result of an agreement, a convention. Thus, the neo-positivist concept differs but very little from the existentialist concept despite the fact that its adherents have in many cases penetrated the foundations of the formal aspects of the information theory, again in the attempt to sanction the pre-cybernetic concept of information complemented by its study and elaboration at the logical semantic level.

In neo-positivism, symbols are almost the only types of reality behind which matter, objective reality, disappears. While the second generation of positivists (Mach, Avenarius) restricted reality to perceptions and impressions today's positivists view this reality in speech, human signs, and study the logical or linguistic aspects of these symbols. In accordance with all this, information is restricted to the sphere of symbols, it is only one of their characteristics (most frequently the meaning of the symbols – the problem of meaning has been termed the basic philosophical question by some representatives of linguistic positivism).

The logical-semantic aspects of information have yet to be elaborated but subjectivistic concepts do, in fact, considerably restrict the extent and forms of such research and prevent the attainment of a uniform comprehensive idea of the interpretation of the substance of information. This can, however, only be achieved if scientists leaning towards positivist conclusions exceed their framework and base their work on spontaneous materialist tradition inherent to cognition in the natural sciences. Neopositivism narrows sharply the subject of philosophy. The same can be said of the effect of the principle of minimalization on the concept of the substance of information – here the substance of information is sought in its individual manifestations in the sphere of human signs and symbols. Numerous fruitful attempts at the use and further elaboration of the concepts of information theory have remained outside the focus of the neo-positivists.

Among the neo-positivists there are some holding views which in an unjustified manner exaggerate the possibilities of the methods of information theory and proclaim them to be something like a universal methodology of science. They now seek the unity of scientific cognition only from the point of view of information (J. Wilkinson, L. Brillouin). The attempt to set the theory of information above all other philosophical concepts is in itself unfruitful and in various concrete situations has proved unsuccessful as well (L. Brillouin). The information theory is not a new universal methodology of the sciences and its concepts from the point of view of contents appear to be more general than philosophical. It can be proved that in any scientific work in which the author consciously refuses to take this or that concrete side in the solution of philosophical problems, he is in fact taking the eclectical side and in most cases only enforces the subjectively idealistic trend as has been the case of L. Brillouin.

The information theory, the same as physics previously, is not the basis for establishing either a uniform language of the sciences or a unified cognition. This is because it only studies one of the properties of mass and one of the categories of material processes. However, the information theory does contribute to the development of integration processes in science, to the synthesis of new knowledge, this unity does not, however, have only a linguistic, mathematic or logical character but is conditional on the material unity of the information processes of communication and control.

This unity does not emanate from the subject as is claimed by the neo-positivists but from the object, it emanates from the object, from the material world which is being studied by the subject. And that is the very reason for which at present the subjectivist interpretation of information is ever more frequently being replaced by objectivist interpretations (e.g. the interpretation of F. Bonsack [1]). The advantage of this interpretation is its broader applicability (the subjectivist interpretations, which is completely justified for a certain category of informational processes also find their application without aspiring to supreme truthfulness).

2. Problem of Objectivity of Information in Current Science

The question of the objectivity of information is linked not only with purely philosophical aspects, with the fundamental question of philosophy, but also with the development of the notion of information in general. For a long time information was conceived as a purely spiritual phenomenon, either as subjective property or as an inter-subjective property. Information was conceived as a statement, a report on an event, transmitted by human speech or by symbols or other means from one individual to another. The view that information is only the property of human consciousness and human communication persists, as has been mentioned above, even today.

The problem of the objectivity of information has become a real problem with the development of the information theory and cybernetics. Thus, for instance, in the information theory (especially in the statistic theory of information) experiments have been carried out to measure the amount of information transmitted by the technical channels of the mass communication media. The notion of information, in this case, was conceived as a phenomenon which destroys, or overcomes, the boundless. It was ascertained in this experiment that mathematical methods based on such a concept of information can measure not only information circulating in the social communications systems but also a number of other processes existing independently of society and human consciousness. What kind of processes are they? First of all those which we term cybernetic. It is known that cybernetics, as was conceived already by N. Wiener, studies the processes of communication and control in live organisms, in society, and in technical installations. Cybernetics has ascertained that control is not feasible without information and therefore all control processes can be considered to be informational processes. From the point of view of cybernetics, processes of communication and control in biological systems also belong to such processes. Biological systems have, however, existed independently of human consciousness, thus cybernetics has discovered a sphere of informational processes in which information is independent of human consciousness.

It is clear that to a certain extent the sphere of technical development is also such a sphere. However, it is not possible to found the concept of the objectivity of information exclusively on the objectivity of technical informational processes. This fact is evident to the opponents of the objectivity of information. Peter Paul Kirschenmann [2] believes that no conclusion on the objectivity of information can be arrived at on the basis of the study of cybernetics alone. He discreetly fails to note the fact that at present cybernetics assigns the property of information even to biological organisms which appeared long before the origin of man. Therefore, speaking of technical systems of control and communication we can state that they comprise and communicate what is a property of human consciousness. We cannot say this is true of information in biological systems. In technical cybernetic installations information exists outside human consciousness but not independent of this consciousness. The technical installation is, on the one hand, an inanimate object, on the other hand, it is a personification of ideas, a projection of the human mind, man's will and work. Yet computer processed information in cybernetic installations has a different form of existence than has information in human organisms.

Biological informational systems are not restricted to communication among living creatures, to the transmission of signals by one individual to another – for this purpose creatures (organisms) use dozens of sound signals, e.g. (the fox has 36, the hen 20, the dolphin 20, etc.). There also exist processes of transmitting information in a different manner such as, for instance, genetic informational processes.

Hereditary information is encoded in the structure of desoxyribonucleic acid (DNA) by the sequence of heterocyclic nucleotides. The genetic code -DNA- is then transferred to the ribonucleic acid molecules (RNA), which in the particularities of their structure (the sequence of amino acids) reflect, reproduce the particularities of DNA. The transfer RNA picks up a certain residue of amino acids and orders it along the messenger RNA where peptide bonds form between amino acids. The process results in the formation of proteins. The sequence of amino-acids in the formed protein, the molecular configuration of the protein transfers the DNA structure and thus also the structure of the biological system passing on the genetic information.

This then is the typical informational process taking place at molecular level in biological systems, a process with a completely material substance. In this way molecular biology together with the theory of information have discovered a sphere of informational processes independent of human consciousness and existing outside it.

Current information theory and especially its application in those branches of science which traditionally study inanimate nature assume that information exists even in inanimate nature. However, to prove the objective character of information, or at least of a part of this information, it is sufficient to ascertain the material character of genetic information and of other types of biological information. The ascertainment of the presence of information in inanimate nature and thus of the general character of information would only add supplementary arguments supporting the objectivity of information, it would enhance the position of the adherents of the material substance of information.

This is comprehended even by the adherents of the concept of information as a purely social phenomenon, as an intersubjective phenomenon. The same P. P. Kirschenmann resolutely opposes all attempts at applying the methods of information theory for the research of inanimate nature and holds that nothing new will come out of them. He denies the very existence of information in inanimate nature [2, p. 185].

This view is not, however, supported by many scientists-physicists, chemists, geologists, geophysicists, geographers, etc. Many of those who have been studying the application of methods of information theory in their respective branch of science hold that information is present in inanimate nature. It has become evident that information in inanimate nature is not identical with cybernetic information, it lacks such properties as meaning, value and certain other properties. In fact, of all the semiotic properties of information there remain only the syntactic ones, i.e., those that manifest the formal characteristics of structure. Structure, form, organization, order is characteristic of all material systems and therefore the methods of information theory are applied in the research thereof.

The results attained by the application of information theory methods of analysis to the study of inanimate nature are not limited to results attained in the application of energy and thermodynamics methods. Naturally, a number of conclusions arrived at are identical but that is natural. It would certainly be peculiar if results attained by the methods of information theory differed or were contradiction to the results attained previously. The principle of unity functions here, the same as it does in physics – in a certain sphere of phenomena the same results are obtained by information theory methods and traditional methods of the natural sciences. In other spheres, however, information theory methods prove more effective, they obtain results "exceeding the possibilities" of the methods applied by thermodynamics. This will become even more evident when we take into consideration the fact that the scope of information theory methods exceeds the scope of probability statistics, including the combinatorial method, the mathematical description of dynamic systems, topology (N. Rashevsky), algorithms (A. N. Kolmogorov), and other methods.

Many natural scientists are convinced that the extent of reality in information is equal to that of space, time, movement, mass, energy, etc., i.e. independent of the perceiving subject. Despite the fact that philosophers are discussing the question of the existence of information in inanimate nature, the author of this paper holds that the future will prove that those scholars were correct who supported the natural scientists in their materialistic concept. The result will, I believe, be the victory of the view that information is the attribute, the property of all matter and that there exist two types of information: non-cybernetic in inanimate nature and cybernetic in live nature and in society.

The recognition that information is a property of matter is a more powerful assertion than the recognition of the objectivity of information. Every attribute of matter is objective and general, it is an integral part of matter in motion, is present in all its manifestations. Regardless of which point of view we take up – whether we hold that information is a property of all matter or of a mere part of matter, we are able to state, on the basis of data available to science, that primary information is material.

If information is an attribute of matter then its existence is infinite, it has never originated and will never perish. Even if information originates at the level of life, as some philosophers are wont to believe, its objective, not its attributive, but its genetically functional character is an important factor for current cybernetics and biology.

In the process of development, systems of ideal information have originated in society. First of all this is everyday information – scientific, aesthetic, religious, legal, political, etc. which are linked with this or that form of social consciousness. All these types of information are aspects of social consciousness reflecting natural and social existence and comprise ideal information. It is ideal information that at the very beginning became the object of the first ideas of information while further development of science showed that the sanctioning of this specific type of information is not justified as it does not correspond to the scientific character of the world. It has become evident that material information is primary, that it exists outside and independently of any consciousness. Ideal information is the reflection of primary information and in this sense it is secondary, it originates at a certain level of development, of the organization of matter.

That is the dialectical materialistic view of the material and ideal information. For some philosophers who adhere to other ideological trends this view seems to be incomprehensible and therefore to them controversial. Thus, for instance, P. P. Kirschenmann notes that Marxist philosophers believe that information exists independently of human consciousness, as an attribute of matter. In this sense it can be considered as objective. Were this so, what then would be ideal information—Kirschenmann puts the question in a critical manner asking: "Can ideal information be objective?" He goes on to claim that this question leads Marxist philosophers into a blind alley and alleges that they are not able to answer his question without violating the principles of dialectical materialism.

P. P. Kirschenmann is wrong, as the correct reply to this question is possible from the position of dialectical materialism. The terms of the question itself are correct only thanks to the fact that in Marxist philosophy the terms "subjective" and "ideal" are not identical in their content. Were they identical Kirschenmann's question would simply be formulated incorrectly as ideal cannot be material and in this sense objective.

Kirschenmann founds his concept on the premise that ideal information is the object of human communication. That is correct, the same as is his statement that the cognition of inanimate objects must be based on these objects themselves. [2, p. 188]. However, from this he arrives at the conclusion that ideal information is objective to the extent to which the transmitted content is dependent of some certain object and is not merely a subjective point of view.

Such a conclusion ensues from a mistake, a misinterpretation of two meanings of the term objective. In dialectical materialist philosophy "objective" is applied in the sense of "material", as a characteristic of objective reality. Kirschenmann applies the term "objective" in a different sense as something inherent to the object of cognition or communication.

Dialectical materialism does not claim that only fragments of objective reality can become the object of cognition, ideas, notions, images, i.e. ideal objects, ideal information—all can become the object of cognition. By becoming the object of cognition, ideal information has not become an objective reality, it remains an object of cognition having an ideal character. Ideal information exists as one that is real, but as one whose existence is not material but ideal.

Here it would be appropriate to note that the term "real" is conceived in a completely different sense than it is by Kirschenmann who views it as a term for all individual objects capable of originating and disappearing in space and in contrary to ideal existence outside of space and time. Dialectical materialism the same as all scientific branches in general does not recognize existence outside of time and space. Therefore the term real is conceived as one describing present or past manifestations of existence and of human consciousness, as conditions of the primary character of existence. The concept of the "real" in this sense differs from the dualist concept which underlies the current philosophical trend known as "realism."

Another question arises, namely, that of the transfer of ideal information, bearing in mind that the process of transmission is material and thus has an objectively real character. The point is that the transmission of ideal information in actual fact takes place by material processes, especially by sound waves, electrical impulses, etc. Symbols are transmitted by these processes and yet the symbol represents the unity of the material carrier of the symbol and its meaning which corresponds to the mutual bond of the objective and sensual meaning. The carrier, the form of the symbol in its entity are the material components, the meaning partly (also the sensual meaning) have an ideal character.

Information in human society, i.e. that information which we term ideal is, in fact, manifested as the unity of the material and the ideal. The carrier of the information is material, the meaning of the symbol is ideal. It is this very bond between the material and the ideal that makes it possible to transmit information (ideal meaning) through material processes. The correlation of the material carrier of information and of meaning becomes the subject of cognition and of communication and remains unchanged in the respective informational control and informational communication processes.

It is evident from what has been pointed out above that the informational communication processes in human society are neither material nor purely ideal, that they can be neither only objective nor only subjective. They are a unity of the material and the ideal, the objective and the subjective. This bond as well as mutual transitions from material to ideal occur because information is manifested as an aspect of reflection as one of its important functional characteristics. Reflection is a characteristic property of matter and consciousness, reflection is that property which is common to these contradictions. Speaking of the relation of information and reflection we have gradually departed from the general aspects of the problem of the objectivity of information to its actual definition.

Above we have discussed information, the object of its existence (existence and consciousness) but we have not made a detailed analysis of this notion. Let us now look into it in more detail and discuss the problem of how the problem of the objectivity of information is reflected in the elaboration of the general definition of the studied notion, founding our analysis on the existence of varied forms ("isotopes") of the notion and its development in current scientific progress.

3. Objectivity of Information and Problem of Definition of the Notion

We have already said above that under the influence of cybernetics and of the information theory, a development has taken place in conceiving information—from the simple report to what is transmitted and what eliminates the boundless. Such a concept of information "works" in those cases where the statistical processes of communication and the transmission of information are studied. It is not, however, the most general concept of information, as there exist communication processes and control processes in which the transmission of information does not take place in the probability statistical manner but is strictly determined.

As an example of cybernetic systems functioning on the basis of the latter principle, let us mention those systems which apply formally logical plans, especially current electronic computers. J. von Neumann terms the information theory, based on formal logic, the strict information theory [3]. We must thus bear in mind that there exist not only probability, statistical informational processes but also improbability, unambiguously determined. The methods of the probability information theory cannot be applied for the study of these processes and that is why various variants of improbability information theories (which we have already mentioned) have been conceived.

The notion of information, used in statistical information theory (i.e. as diminshing uncertainty) is useless for the research of unambiguously determined communication processes. In the transmission of information uncertainty is not overcome but information simply moves from one object (the source of information) to another object (the addressee of the information). What then moves and what is transmitted by the communication channels in unambiguously determined processes?

The reply to this question was given by N. Rashevsky in 1955 and independently of him by W. R. Ashby in 1956. The former studied the topological informational properties of biological systems and arrived at the assumption that the multitude of information contained in them is determined by the multitude of topological differences, i.e. by the topological variety. Rashevsky used the methods of probability topological information theory and yet the approach which he suggests is one based on the notion of topological variety.

W. R. Ashby in his book *An Introduction to Cybernetics* [4] made, we believe, a successful attempt at interpreting cybernetics and the information theory on the basis of diversity, variety. According to Ashby the transmission of information is equivalent to the notion of variety. He built up this concept of variety only on the basis of the statistical information theory at a time when the non-statistical variants of this theory had not yet been developed – he assumed that the set with probability is equivalent to the set with variety. It later became evident that the concept of information as being one of the transmission of variety is methodologically fruitful even for the improbability non-statistical information theories and their applications – and this fact has been proved in the works of Marxist philosophers [5; 6]. In this manner the concept of information as variety became widespread and the transmission of information was conceived as the transmission of variety.

Information is not, however, identical with variety. One term cannot be replaced by another and have the same meaning. The notion of information is conceived not so much as variety but as the transmission thereof. But the transmission of information is only one of the forms of the movement of information, next to it there exist such forms as storage, transformation, reception; what then is it that brings these forms of the existence of information close to one another? What is it that differentiates information from variety?

In their works Marxist philosophers have shown that there exists a notion which links all forms of the movement of information, namely that of the reflection as the general property of all matter. From this point of view information is not variety but a reflection of variety or the variety of reflection.

The notion of reflection as the general property of matter was introduced by V. I. Lenin in 1908. He then said that "in the foundations of the very edifice of matter" we can assume "the existence of an ability similar to sensations" [7, pp. 39–40]. V. I. Lenin assumed that reflection is the property of one material object to reproduce the content of another object as a result of their mutual correlation. In this sense reflection is similar to sensations which are inherent to highly organized matter alone.

The reflection of variety is thus one certain aspect of the reproduction of the content of one object in another object and information is thus only an aspect of reflection. The notion of information has thus united in it two concepts: variety and reflection. This, the most general concept of information which does not contradict any of the existing information theories is at the same time broad enough to allow some future variants of the information theory to originate. More and more philosophers have recently tended to agree with this concept of the information theory and they have been joined by an increasing number of experts in the field of cybernetics, information theory and its application.

If we interpret the notion of information as variety of reflection it is evident that this definition also solves the problem of the objectivity or information. The individual elements—notions which comprise the notion of information (reflection and variety) are objective, they are inherent to all matter. Reflection is not only a subjective reflection of the outside world (i.e. the gnoseological notion of reflection) but the general property of matter, independent and existing independently outside the consciousness of man. V. I. Lenin in his thesis on the existence of reflection as the general property of matter established as objective something that had previously been conceived as something subjective. A similar phenomenon has now appeared insofar as the notion of information is concerned. The difference is only in that reflection used to be linked with the processes of cognition while the notion of information is linked with the communication processes. Therefore, the current interpretation of the notion of information on the basis of the notion of reflection emphasizes even more the need of the objectification of the content of the studied notion. At the same time it is a contribution to the synthesis of the notions of cognition and communication in a more general notion.

The notion of variety expresses the summation of differences in the object. The notions of multiformity, difference, diversity, etc., exist and have a similar content. They exist objectively, are inherent to all material systems and from this point of view are the attribute of matter. Insofar as information is concerned, it represents the actually existing basic link between two existing attributes of matter and as such it is also the objective property of matter, existing outside and independently of consciousness. (The two existing attributes of matter being reflection and variety). From this point of view, information is also an attribute of matter, as being the inter-relation of two attributes of matter it is also an attribute.

Thus, the definition of information as being the reflection of variety presumes the objectivity of information and, moreover, it presumes its attributive character. The latter is not, however, indispensable for proving the objectivity of information. The other above-mentioned point of view held by certain Marxist philosophers, namely, that information originates at the level of life, is also based on a materialistic foundation. The discussion focuses on the object of information alone and does not rise from basic philosophical differences but from blank spaces in the current level of scientific knowledge and this is a problem which not only philosophers but also scientists and scholars working in other scientific branches have to face and solve, as scientific progress bears with it new and as yet unsolved problems.

Naturally, the scope of every notion is related in a certain way with its content (in formal logics – the law of inverse proportion of the scope and content of the notion). Therefore, the content of the notion of information is inter-related with the object sphere of the existence of information. No matter what the position of the Marxist philosopher on this particular aspect should be, he has always presumed the objectivity of information.

In those cases where the scope of the definition was narrower (and in such cases we believe that only some forms of the notion of information were concerned), the Marxist philosophers never departed from the view that information exists outside and independently of man's consciousness.

It is true that the author of this thesis does not agree with the narrower concepts presented by some scholars, yet in no case is the point of disagreement linked with the aspect of the objectivity of information. The fact that the author gives priority to the broader concepts ensues from the trend towards objectification which is manifested mainly in the object sphere of the notion of information. The concept that information is a purely spiritual phenomenon has gradually been objectified and this has, in our view, strikingly extended the scope of the objects reflected by information. This broader concept today appears more justified from the philosophical point of view but is also used as the basis for current theories in the field of cybernetics and of the information theory elaborated in the USSR by Academician B. N. Petrov and his cooperator and followers [8].

The scholars who adhere to this scientific trend conceive information as objective to the same extent as are physical properties and characteristics in general (energy, mass, space, time, etc.) They also uphold the concept of variety and are developing an original mathematical system to study the processes of communication and control. While previously the processes of communication and control had been studied separately from the strictly formal mathematical point of view – as information theory and the theory of automatic control – the concept of an informational control theory establishes the basis for a common language enabling the study of both communication and control processes.

This concept is founded on the broad concept of information, namely on the informational analysis of control processes. At the same time this makes it possible to establish cybernetics not only as a scientific trend based on concepts at the content level but as one uniform branch of the sciences with its own specific mathematical system. That, too, was the mathematical "vision" of cybernetics, the dream of N. Wiener. It is founded on the objectivity of information, on the concept that all matter is "saturated" with information and that cybernetic systems only exploit and reflect the variety of material systems which surround them.

The informational approach to cybernetics, based on the concept of objectivity and of the general character of information is now yielding its positive results. Information is that factor which unites two relatively different cybernetic processes, namely communication and control. It is that factor which allows cybernetics to reach the status of a uniform comprehensive scientific branch such as is physics. The establishment of physics as a comprehensive scientific branch was brought about by the enforcement of the concept of the objectivity of energy –

previously energy and power also had a subjective flavour to them—and by the enforcement of the concept of the general character of the energy method. In cybernetics the analogical part is taken by the notion of information and with it various inter-related information theory methods of the study of processes of communication and control.

References

[1] F. Bonsack: Pour une interpretation objectiviste de la Théorie de l'information. Dialectica *16* (1962), no. 1—4.

[2] P. P. Kirschenmann: Information and Reflection. On some Problems of Cybernetics and How Contemporary Dialectical Materialism Copes with Them. Reidel, Dordrecht 1970.

[3] J. von Neumann: Theory of Self-Reproducing Automata. University of Illinois Press Urbana and London 1966.

[4] W. Ross Ashby: An Introduction to Cybernetics. Chapman & Hall, London 1956.

[5] J. Zeman: Poznání a informace. NČSAV, Praha 1962.

[6] A. D. Ursul: Information — Eine philosophische Studie. Dietz, Berlin 1970.

[7] В. И. Ленин: Полное собрание сочинений, т. 18 (V. I. Lenin: Collected Works, Vol. 18). Москва 1961.

[8] Б. Н. Петров, В. В. Петров, Г. М. Уланов, В. М. Агеев, А. В. Запорожец, И. Д. Кочубиевский, В. П. Май, А. С. Усков: Начала информационной теории управления (B. N. Petrov, V. V. Petrov, G. M. Ulanov, V. M. Ageyev, A. V. Zaporozhec, I. D. Kochubiyevskiy, V. P. May, A. S. Uskov: Principles of Informational Theory of Control). In: Итоги науки и технили. Серия: Автоматика у радиоэлектротехника. Выпуск: Техническая кибернетика 1968. ВИНИТИ, Москва 1970; also in: Техническая кибернетика 1969. ВИНИТИ, Москва 1971.

INFORMATION, SIGN, IMAGE

Herbert Hörz

The views expressed in this paper deal only with a few philosophic aspects of the information concept. The role of information as a sign will be treated as a specific process of human reflection which produces images of objective reality. In this process the concept of information appears in different connexions. First, information is considered as a part of reflection, which is a general property of matter. Reflection is to be understood as the production of traces by a material process, which are left behind on a system and which reflect part of the structure of this process. In inorganic nature, effect is followed by counter-effect and traces can be extinguished by new effects. Information is understood as a measure of order, as a structure suitable for controlling and regulating the above processes. However, this statement is problematic in the case of inorganic interaction. The traces, which appear with it, may cause system-particular reactions. Some authors therefore consider it wrong to speak about information already of this level. They consider the above processes only the prerequisites for later information processes. This point of view has much in itself and it would eliminate many an anthropocentrism.

In information theory information is connected with yes/no decisions. If an objective state or sign has only two configurations, then the appearance of one or the other is a decision, which is characterized by a bit as a measuring unit. Wiener defines information as an alternative decision by which he understands "the choice between two equally probable, simple alternatives of which one or the other must occur, e.g., tossing a coin 'heads' or 'tails' " [1, p. 87]. Information proceeds from the sender to the receiver, it is coded by the sender and decoded by the receiver. For that purpose it is necessary to know the information unequivocally as well as the code. In this sense, when using Morse code, messages are transmitted by means of dots and dashes in a known cipher. If we apply this concept to inorganic nature, anthropocentrisms will originate, because a stone makes no decision but evenso behaves lawfully. However, the whole proposition can be objectified by taking into consideration the field of objective possibility and the incidental realization of possibilities, contained in the law, of influencing an inorganic system of elements, i.e. an object, by a material process and if thereby traces are left behind on the system which contains part of the structure of this process than we have a prerequisite of reflection a pre-form of reflection, or reflection. The structure of the process contained in the system may be considered as information about this process. However, as the inorganic system does not react system-particularly to traces, but only directly to effect, this is a potential information which becomes an actual information for the cognizant man if he draws conclusions from traces about effect, perhaps from mountains about the distribution of glaciers in the ice-age. Because of the very character of such information as potential, some scientists do not speak about information here. In my opinion, however, the problem is not a matter of definitions, it seems to lie in the fact that these forms or pre-forms of reflection are acknowledged as existing, because otherwise there would arise difficulties in explaining life and consciousness as a developmental process.

Secondly, the term of information processes is in common use in molecular biology and among philosophers when talking about living beings. Here the problems of anthropocentrism do not seem to be so pronounced, because living beings adapt themselves actively to the environment, they establish their own system which is dependant on information from the environment in order to be able to remain adapted, and which corresponds to information by system-particular reactions.

While in inorganic nature it is possible, to speak about pre-forms of information or about potential information, living beings dispose of an ever developing apparatus for receiving signals (sign) from the outside world to which they react system-particularly.

The worke-up information is condensed to a program of self-regulation by which reflection is again affected. Although it is impossible to deal in detail with these cybernetic regulation processes, the importance of information as signals from the outside world or of the organism itself which lead to a specific reaction of the organism, may have become clear. Therefore the question is not whether this process is expressed anthropomorphously or not, but that a new complicated information—and reaction-mechanism originates with the development of living beings. This mechanism works up biological stimuli, recognizes signals in connexion with biological stimuli, developes possibilities of reaction towards them and thus acquires an information-reaction-apparatus which serves an active adaptation to the environment.

Primitive living beings give only a lesser specified overall reaction to biological stimuli. Signals, connected with biological stimuli, serve at higher levels for the formation of conditioned reflexes, and in the thinking behaviour of higher animals there originate pre-forms of information combination for establishing a program of activity.

Thirdly, for this reason, information as a part of a specifically human reflection, of the image processes, is of great importance. Proceeding from Helmholtz's empiricism, some theoreticians, among them Plekhanov, denied however the role of our cognition as image. Cognition was to be only a sign, but not an image. Lenin argued against this opinion, because signs may be arbitrary, and are not necessarily linked with the reflected process, and therefore this opinion lends itself to subjective and agnostic interpretations. In contrast to idealism Lenin stressed the primacy of matter, against agnosticism the reflex character of cognition. Helmholtz had turned in the case of sensations against the expression "image" (Abbild), because there is no identity of the reflected and the image. However, he had always emphasized that signs mediated to us messages about the outside world.

The identity of the image and the reflected is not asserted even in the dialectic – materialist theory of reflection. However, it is stressed that by means of material signals and their ideal meaning one recognizes and expresses in theories the lawful behaviour of objects. These theories lead via rational processing to instructions for acting, which are possible reactions to environmental effects, but they also serve the creation of conditions for the realization of self-set goals, which serve the production of conditions of human existence and comprise not only material products, but also the entire, intelectual and cultural sphere.

An important difference between information, sign and image is to be seen, therefore, in the different significance for inorganic nature, life and human society. In inorganic nature the system which is affected changes, and there originate traces of effect. Information contained in these traces remains potential until man concludes from it upon effect. Information here is not this effect itself, but only the traces left behind, because the system does not react to effects system-particularly, as if it had received a signal, but directly to the energetic and material effects. Since this effect possesses no signal character, as it does not appear as a sign for other connexions, some authors do not speak about information even in this case. In my opinion, it is correct to speak about potential information here, because man can use traces in changed systems for conclusions concerning effect. In a living being the reaction is different. In this case, effect as a biological stimulus is associated with signals of biological stimulus. The living being can react in a system-particular way to a signal or sign by which it is informed about the presence of biological stimulus. In the case of inborn, unconditioned reflexes it reacts at once, while in the case of conditioned reflexes an established reaction developes only gradually. Here the signal as an information changes the biological system in certain elements and subsystems until the corresponding stimulus-reaction-mechanism is established. Thus a living being reacts to the signal or sign at once, if the information character of the sign is inborn, and it learns the relation of signal and biological stimulus, forms of it a temporary connexion, if such relations appear regularly. Such experiences can be stored and in the case of higher developed animals a combination of activities is brought about which corresponds to the assimilation of different informations gained by experience. Therefore, signs or signals can be considered as coded information which is being decoded by the animal in the process of learning. At the same time, signals and signs arc limited only to the biological life process of the animal. Signs have no meaning to the animal without a corresponding biological stimulus. If sign and stimulus appear no longer together, then the contemporary connexion of both will dissolve gradually. The decoding of signs is therefore, a material process and it does not permit the origin of any ideal image of objective reality, which could be combined with other images later on, and which thereby would lead to new images. The combination of ideal objects to an ideal program of material activity appears only with man. As far as man deals with the information process of animals, he himself acquires information which he works up theoretically.

The information process differs in man's case principally from the informatiom process in the case of animals. Thus we speak of a specifically human reflection process in which ideal images of objective reality originate. However, also in the human reflection process, information and image do not coincide. Man reacts on the basis of his developed signal system, in which also the words as the second signals ar included, to a great variety of informations of the outside world and of one's own state in the sense of the material information-reaction-mechanism.

Information, which appears as signals whose relations to the signaled process are known by experience, by theoretical acquisition, or by their connection to known signals, will not be consciously decoded, but will be answered immediately by reactions on the basis of the already existing contemporary connections. Here, the material information-reaction-mechanism of man differs from that of animals. Man namely is able to generalize in such a manner that he can react in the same way to materially different signals, which signal the same structure. He uses in particular words, which have been endowed with contents by theoretical work, i.e. images, because they represent a unity of material process (signal, sign) and ideal meaning also as the second signals in the material information-reaction-mechanism. Let us think of the reflex behaviour of a trained motorist who does not work up the information received in a moment of danger consciously, but on the basis of experience reacts to signals immediately in the material stimulus-reaction-mechanism. Information from objective reality has here, as a sign, a signal character for situations to which there exist established ways of reaction, which man need not think over thoroughly at first. Thus the lights on red forces the motorist immediately to release the reaction of braking process without being obliged to become aware at first of the connexion between traffic rules and his behaviour and then to think about how one must brake.

Up to now "signal" and "sign" have been used as synonyms, because signal is transmitted as a sign. The buzzing of a midge is a sign for a spider to which it reacts in the same way as when a tuning-fork resounds near its web. The sign is a signal for the presence of a biological stimulus. In the same way the traffic signs have to be observed by the motorist as signals for certain situations which require certain reactions. The concept "sign" is sometimes limited to language and the word is considered as a sign which possesses meaning. This limitation seems unfortunate, when it leads, as in the case of Zhukov, to distingushing signal as a material carrier of information from sign, whereby the latter is to be a material expression of social information and is not to be linked to the transmisson of energy [2, p. 139]. We shall return to the distinction between signal and sign later on. First, however, let us make a remark on the transmission of energy by word, which is considered by Zhukov as a sign only. One understands by language a system of meaningful sound signs which can be also expressed secondarily by letters, Braille, Morse signs, etc., which serve the understanding within a community. Language as a means of understanding is, however, linked to consumption of energy. The word exists only as a written, spoken word. However, writing and speaking consume energy. One could presume that, as long as the word is not expressed, no energy would be needed. On the other hand, investigations concerning the so called inner speaking have proved that—in the case of strenuous thinking activity—there occurs sensibility of vocal muscles as well as corresponding electro-physiological impulses. Zhukov believes that the transmission of information by word takes place without any consumption of energy. However, this is not so. The word must be recieved and assimilated, which consumes energy. While the material signal affects the sense organs directly and thus transmits information with little energy, the ideal signal must effect the sense organs as well. However, this need not be direct. A book or a record can remain unused for a long time. However, effect itself requires energy. A silent word or an invisible letter are no signals. While, of course, in the case of a material signal energy and transmission of information form a unity, in the case of an ideal signal not the material effect with its structure acts as information, but the ideal contents of material process possesses a signal character, it is a sign of meaning.

When gaining its first experience, a child sees objects and hears words for them. The linguistic share of this sign, which is transmitted by means of energy, is a signal for the share of cognition. Every increasing experience, linked to further learning of linguistic signals, leads to generalization, and the word becomes a signal for clases of objects. The word becomes a concept, i.e. a summary of social and individual experience with objects denoted by the word. In Hegel's words, the known then becomes the recognized. The energy, needed for transmitting the word, does not change, but rather the share of cognition which is linked to it. It is therefore interesting that the linguistic share is not essential, as long as the sensual formation of experience prevails, as it is shown by clinical experience with aphasia, the lack of stability of speaking, but that it is advantageous for the work of memory and for the ability of generalizing. In concept formation, words and word groups are, however, the material substratum for the formation of ideal objects and for the preservations of cognition in memory. Simultaneously, the concept contens change with the process of cognition and the same sign becomes a signal for more comprehensive knowledge and cognition. Consequently, energy is consumed both in the formation of words, in their articulation and writing, in reception and assimilation. However, the same transmission of energy necessary for certain words may lead, in connexion with energy consumed for thinking processes with sender and receiver, to greater information than before. In fact, this is a difference between the material signal, which mediates information by its structure, and the ideal signal, which possesses a material carrier and consumes energy for transmission, while the content of the signal is, however, not linked to the material structure, but can possess – depending on the sender and the receiver – a greater or lesser share of cognition. This may have caused Zhukov to omit entirely the transmission of energy as unessential.

Accordingly the word now has a double function in the human process of cognition. Once it is a material signal in the sense of Pavlov's second signal system and it leads to releasing a stimulus-reaction-mechanism. Here, social and personal experience enter into this signal, which is connected with the other meaning of the word as a sign for a concept, for an image. In the first case the word signals an objective situation to which one should react, and if there exists the stimulus-reaction-mechanism for it, one reacts in a corresponding way to the information about the existence of the situation which is transmitted by the word. Thus a trained soldier can react by contitioned reflexes to commands, which would demand long reflections and possibly cause wrong behaviour in the case of a man who does not perform military service. He tries to become aware of the objective situation concerning the image character of the concept (concept system), which is signaled by the word (command). The utilization of this double character of word as a signal for objective situations and as a sign for images (concepts) makes also Zhukov's differentiation of sign and signal problematic. The word as a sign is both a material carrier of information – which would be denoted by Zhukov as a signal – and a sign for a concept – which can be understood in accordance with him as a material expression of social information. The fact that in the case of a sign as a material carrier of information the stimulus-reaction-mechanism is concerned, which is released by material signals, is also proved by the fact that words become such signals for animals. The new quality in the case of man lies, however, in his capability of distingushing between similarly sounding words in their contents and thus he utilizes jointly the second side of word as a sign for concepts. From this point I see no need to distinguish between sign and signal, but I do consider it necessary to differentiate between the meaning of a sign as a material signal for objective situations and as an ideal signal for images. Even for an animal the sensation of a colour, of a certain tone is a sign, which appears as a signal for objetive situations, but here the signs do not possess the double character of material and ideal signal.

Man unconsciously works up information received from objective reality, with the aid of concept formation, generalization and creative ability of thinking. The part which is determined by unconditioned and conditioned reflexes and which is also based on the material signal character of word, releases corresponding reaction mechanism. As not everything has to pass via man's theoretical consciousness, he can select information which he transforms into theories on objective reality, including meta-theories of different semantic levels in order to gain an ideal image of objective reality, which leads to an ideal program of further activity and – converted into directions for acting – releases reactions. Informations are thus transformed into images in a specifically human reflection processes. Here there exist different relations between the ideal and material signal character of word. As seen before, the rational processing of information is led to images in theories up to directions for acting. Thus, by means of permanent repetition, the process of assimilation of information leading to theories, directions for acting up to reaction can be shortened, if the ideal signal character of word, which signals the image is neglected and only the material signal character of word acts and releases reactions. Vice versa, however, a reaction linked for long time to the material signal character can be checked by theoretical analysis and it can be established anew. A wrong or unrational manipulation in work process, an unhealthy posture of a pupil at school, etc. can be changed gradually only by permanent explanation and conscious utilization of arguments for new ways of acting. In this sphere of interrelationship of material and ideal signal character of words there are many problems yet unsolved, which concern subconsciousness, preconsciousness, etc. For instance, one may react to the material signal character, subconsciously, although the direction for acting would have been differently had there been a conscious analysis of information and working-up into images of the objective situation. The very ideal signal character of word, concerning the word as a sign for the whole of experience with denoted objects, which are summarized in the concept, express the new quality of the specifically human reflection process. The experience cannot always be divided into clear yes/no decisions. It encompasses the acquired scientific cognition, personal experience and certain presentiments, unclear conceptions, etc. The exact image should be herewith determined by yes/no decisions, while – in order to speak with Weizsäcker – in this case a hardened point of an unhardened matter is concerned [3, p. 60]. Let us take, for instance, the term "elementary particles structure" (Elementarteilchenstruktur) and let us try to determine the experience we have made with it, we shall arrive at the sphere of hypotheses in this way. Thus the search for fundamental particles, the acceptance of an elementary length, the transmittance of nuclear models to elementary particles structure, the bootstrap-hypothesis for strongly interacting particles, etc., are connected with this term. There exists experience for the elementary particles structure. Essentially, it can be also formulated as information which is unequivocal and whose contents are yes/no decisions. The best way for doing so is by a clear description of experimental results. However, there does not yet exist a theoretical generalization which explains experimental results, discloses their inner connexion, formulates laws, etc. The term "elementary particles structure" is herewith already a concept which comprises experience. This concept is an image of objective reality, because it codes objectively real processes in a specifically human way and grasps them as cognition; man is able – according to his experience, i.e. by means of his knowledge of the code – to convert the theses and statements which correspond to this concept into directions for acting, and to react correspondingly. However, this conception as an image is not to be reduced to information. Thus the specifically human reflection processes differ from the information process in the way that information in inorganic nature is only a potential formation of traces of the effect, without information existing as signs or signals for system-particular ways of behaviour towards objective situations of the environment or of the system itself. In the case of living beings information has a material signal character, as a sign it elicits the responses of the system, which are – on the basis of the laws of the system – responses specific to the signaled objective situation. Man assumes the stimulus-reaction-mechanism on a high level, i.e. embedded in the reflection process, and he posseses the word as a material and ideal signal. Here, however, the difference between information, sign and image in the human reflection process becomes interesting.

First the words function as a sign for objective situations and require corresponding reactions, released through the word as a material signal. This can be a matter of a behaviour practiced again and again, based on discipline, respect to teachers, instructors, etc., and appears as a word-reaction-mechanism. One's own experience may also have formed this process. Here, the word was a sign for the ideal image of an objective situation, which was analysed, theoretically grasped and conducted towards directions for acting. The reactions were practised until the word became a signal of the objective situation and gave the sign for the corresponding realization. Therefore, here the word acts as information.

Secondly, however, the human reflection process follows via consciousness by creating the products of cognition as images of objective reality. These images are concepts and theories whose signs are words and word combinations, and these signs signal the objective contents, i.e. objects and processes (objective truth) existing outside consciousness and independently of it, as well as the experience with objects and processes (relative truth). Therefore, images are worked-up informations, which themselves again possess information character, while unequivocality cannot be reached with all for the whole of concepts and theories and cannot by reduced to yes/no decisions.

Thirdly, however, there also originate in the reflection process hypotheses, models, etc., through the working-up of information. New ideal objects are created, whose reflection character is not yet known. Images, such as the concept "elementary particles structure", cover objectively real processes and the experience with them; however, they cannot yet be defined univocally. Possible ways of thought are investigated. Here the actual *cognition process* is concerned, in which images are used and new ones originate. It is the ideal reflection of material and ideal processes and includes not only information, although it must repeatedly lead to information.

Fourthly, reflection is used for directions for acting. The reflection process leads to ideal model of objective situations, which can prove its significance only via directed responses (experiment, practice). Here reactions are therefore determined not only by information, but also by the entirety of images, i.e. by hypotheses, presentiments, etc. which are worked up to directions for acting to which there are encompassed information, hypothetical cognition, norms, set goals, and problems. The reaction serves the change of environment and the receipt of new information for working-up, in order to be able to set adequate images as a basis of directions for action. The information process therefore does not fully include the reflection process in the case of an ideal working-up of information and of course not the release of reactions. The reaction itself is not only the result of information.

Fifthly, one must also draw attention still to another component of the reflection process. The reaction is controlled not only by information and reflections, but also by emotions. Emotions express the abundance of man's relations to the environment. They are man's psychical responses to the effects of the environment and are conditioned by the state of the organism. They are a part of the reflection process. Without any doubt the education of emotions pays a great role. However, it would be wrong to consider emotional reactions as the lowest stage of psychic reactions. Science and personal experience show that emotions are reactions to the effects of the outside world which are worked up by man in a specific way. Emotions depend on information. They express in a specific way, as pleasure, happiness, anger, affected behaviour, etc., the cognition of objetive situations. Emotional reactions are thus released by certain information, but they are not there immediate outcome.

The specifically human form of gaining information is therefore a part of the human reflection process. Information is a basis for images which arise in the process of assimilation of the information and which are themselves again determined so exactly that they can be returned as information. However, image is more comprehensive than information. It encompasses experiences which still need to be determined as information. It expresses objective dialectics in subjective dialectics. Via directions for acting, it controls reactions which are however accented also by emotions. These relations show the complexity of the reflection process and the necessity of investigating still further the relation of image and information. The relation of signals and signs in the process of reflection has been pointed out. Through it also the much discussed problem of the identity or non-identity of object and image are solved. Signals and signs possess the function of mediation between object and image in human cognition. We do not consider here the material signals which, as information, elicit responses immediately, because they are signs of the existence of an objective situation, which requires a certain reaction, but only the images. A certain objective situation leads to a number of effects on man, which are transferred further as signs or signals, according to the human organism and its cognition apparatus. These signs are coded messages from the outside world.

Man has mastered this code during his long development; he learns to understand it better by modelling for himself the objective situation on the basis of signs, by setting up directions for acting, and by reacting. The result of the reaction permits conclusions to be drawn concerning the ideal mode of the objective situation. With the help of practical activity man checks whether he knows the code and he decodes signs and signals. The process of decoding the signs, transmitted by energy, is linked to the experience of man hitherto. If inexperienced, he cannot interpret the abundance of signs, he disregards other signs and thus a limited image of reality is produced. The process of cognition is not only the reception of signals and the decoding of signs, but a rational assimilation of signals from the outside world in the connexion with the possibilities of thinking. Creative thinking proves true only in the cognition of objective laws, which – coupled with logically possible signs – leads to an ideal model of objects to be realized and which are realized, via a program. This is what is accomplished by the reflection character of our cognition. The image as an ideal model of real relations permits, via the totality of directions for acting, as a program, not only purposeful actions for the purpose of adaptation to the environment, but also active change of the environment, the creation of an artificial outside world, where the conditions of man's existence can be produced in an always ampler and better way. Seen from this viewpoint, Lenin's criticism of semiotics is more than topical. Certainly, we do not have any images of objects in our heads but we have been able to establish with the aid of signals from the outside world a meaningful system of signs, which represents in its inner connexion a verifiable ideal model of objective relations with whose aid one can react appropriately and change purposefully. This model is further developed by new signals and reactions, new signs join it, the meaning of known signs is better recognized, etc. The sign is therefore a signal for the image and not the image itself. The signs alone would indeed transmit information and elicit reactions, if the connexion of sign and reaction has been established as a conditioned reflex on the basis of experience, but they would not permit an ideal image. Only as far the sign does not only signal an objective situation, but also the existence of the concept as an image, we can speak of a specifically human reflection process in which signs signaling images are decoded, combined, stored and transmitted further.

The presented problems and their solutions have shown how important it is to elaborate further the theory of cognition and to take into consideration the different aspects of cognition.

References

[1] N. Wiener: Kybernetik. Hamburg 1968.
[2] Н. И. Жуков: Информация ((N. I. Zhukov: Information). Изд. Минского Университета, Минск 1971.
[3] C. F. v. Weizsäcker: Die Einheit der Natur. Hauser, München 1971.

PHILOSOPHICAL PROBLEMS IN THERMODYNAMICS*

Benjamin Gal-Or

I. Introduction

At the beginning of his studies in natural sciences a student sincerely believes that such fundamental concepts as time, entropy, temperature and irreversibility are well defined and well understood by the experts, or at the least, by the well-known authorities in thermodynamics, statistical mechanics and related subjects. It is, however, highly perplexing and embarrassing for a thoughtful student to slowly discover later, that one after another, the experts and the well-known authorities have never been really certain about the origin and meaning of the very fundamental concepts embedded in their own theories. Furthermore, often the authorities are in serious disagreement with each other (and with other views expressed by their colleagues), as to the proper definitions, origin, validity, and meaning of the aforementioned concepts. Sooner or later a person seeking a more consistent and universal answer to these questions finds that any respective answer requires endless (but seamless) links between philosophy of science and the more earthly problems of classical thermodynamics, statistical mechanics, electromagnetism, relativity, and even cosmology!

Such a searching course into the open and unknown problems of today, requires the questioning, re-examination, screening, and, at times, radical rejection of "accepted" or "established" theories of past or present authorities. It criticizes and reprobates any traditional teaching, which eventually leads to inconsistencies or paradoxes, no matter how sacrosanct the "theorem" may be. It defends and preserves no argument which does not, by its own logic, consistency and universality, defend and preserve itself. Such was always the course situated under the crisis of every science including thermodynamics. The crisis, as well as the problems to be treated in this report, manifest themselves most clearly by new attempts to provide answers to such fundamental questions as:

1. Is the origin of irreversibility and the law of increase of entropy *deduced* from the mathe-

* The paper is the second part of a previous essay "Entropy, Fallacy, and the origin of Irreversibility" which was accorded the Gold Medal of the New York Academy of Sciences (small Anall. of the N. Y. Acad. Sci. March, 1972).

matics of statistical mechanics applied to a large number of particles or had it been inserted into the theory without a due declaration before the first equation was actually written?

2. How may one reconcile the paradoxical aspect according to which, on one hand, observations demonstrate that throughout space accessible to our observations, the properties of nature bear no resemblance to those of an equilibrium system (e.g. the red shifts, Hubble's law, nuclear reactions, temperature gradients in the universe, and the nonequilibrium expansion of radiation energy into cold, non-reflecting expanding space), while, on the other hand, statistical thermodynamics claims that *any finite region* of the universe, no matter how large, should have a finite relaxation time and *should thus be in an equilibrium state* [13]?

3. Is the *origin* of irreversibility in nature local or cosmological? In other words, are there physical links between the expansion of the universe, the irreversible flow of radiation from all galactic bodies into cold, non-reflecting space, and the entropic, information, electromagnetic, biological, and statistical arrows of time [2]?

4. What is the basic nature of the somewhat mysterious time coordinate system in which the very physical laws are embedded?

5. If practically all (see, however, below) *microscopic* equations of motion are invariant under time inversion why do most natural processes operate only in one time direction?

6. Is the recently discovered *microscopic irreversibility* in the $K^\circ - \bar{K}^\circ$ decay of a neutral kaon [12] linked by a one-to-one relation with the cosmological asymmetry? Or, perhaps, must its ramification lead to a new look into the very foundations of statistical mechanics?

7. Is entropy an *essential* element to thermodynamics? (e.g. does a unique entropy value exist in irreversible processes [1]? Or can a consistent *entropy-free thermodynamics* be compatible with other concepts, conclusions and observations related to small as well as large-scale systems?)

8. Why should there be any time assymetry at all? and why are all time asymmetries that are in evidence (including the thermodynamic, electromagnetic [16], microscopic [12], biological and cosmological [25] ones) consistent with each other in pointing to the so-called positive time direction?

Obviously, such questions, and many related ones, are so fundamental that no thoughtful person can claim to have given them no consideration. It may also be noted that many of these problems formed the frame of research for the greatest minds in the past. Today, however, the student is trained specifically *not* to ask them. *Consequently, awkward or incorrect points of departure have now an increasing chance to survive for a long time.* A small, but a steadily growing number of scientists, have, therefore, come recently to believe that such basic questions, which are related to the foundations, and the very core of most of our theories, should be re-examined periodically in publications and in international meetings till a more unified approach is established. Such rationale catalyzed in recent years the activities in at least two major international conferences. The first, and probably the most important one, was organized by T. Gold and took place in 1963 at Cornell University with the participation of some leading scientists. The informal discussions were edited later by Gold [4].

The second conference was held at the University of Pittsburgh in 1969 (its proceeding *A Critical Review of Thermodynamics* [1] were edited by Stuart, Gal-Or and Brainard). Some of the leading thermodynamicists provided in this conferene reviews and criticisms of the philosophy of present-day thermodynamic theories. It is a remarkable fact, however, that no agreement was reached in any of these conferences about the origin of irreversibility and time asymmetries in nature.

While the answers may not be known with certainty for some time to come the discrepancy of opinions among the participants is, perhaps, symptomatic to the increasing crisis at the very foundations. Some of the roots of these problems may be found in some "established" traditional thinking and teaching which, in turn, carry their momentum from the impact of a few authorities in thermodynamics, statistical mechanics, relativity, etc.

While these conferences show that during the last decades there has been a remarkable activity in statistical mechanics and irreversible (classical and relativistic) thermodynamics, most of the basic problems involved, have largely remained unsolved. Critical reviews of the yet unsolved problems are available elsewhere [1-3; 19] and would, therefore, not be repeated here. I only stress that in spite of an extremely diversified spectrum of opinions, one may roughly distinguish here among *three main schools of thought*. These are:

1. The traditional axiomatic (classical, special and general relativistic) thermodynamics with some newer and more refined modifications, which, however, supply no new answer to the existing problems embedded at the very foundations [1–3].

2. The more "established" statistical school which generates irreversibility, entropy, information, and electromagnetic and thermodynamic time asymmetries by an *a priori* deliberate *choice* of an *initial condition* imposed on time-symmetric probability theory, so as to fit the *local* behavior of the universe and the classical "second law" of thermodynamics [1-3] (see also below).

3. The astrophysical revolutionary school which deduces the *origin* of irreversibility, electromagnetic and thermodynamic time asymmetries from the large-scale non-equilibrium dynamics of cold, non-reflecting, expanding space which, in turn, is intimately related to initial conditions imposed on time-symmetrical field equations [5–7; 15] (see also Fig. 1).

Due to the fact that the basic thoughts underlying the first two schools are, by far, more "established", known, supported and "accepted", as well as due to space restrictions, the present report is concerned mainly with the last school of thought. However, in the next section, I shall briefly outline the question of initial conditions in the statistical school which play also a central role in the fundamental concepts of the last school. Disagreement, as I shall show, is inherently about the *status* and *origin* of "proper" initial conditions.

II. The Failure of Statistical Mechanics to Deduce Irreversibility

This section deals with questions 1 and 2 of the introduction.

The origin and deduction of irreversibility in (classical or quantal) statistical mechanics has raised many questions. For instance, it has been felt by many eminent scientists that the theories of statistical mechanics and thermodynamics contain few "paradoxes", notably the well-known Loschmidt [36] and Zermelo [37] "paradoxes" [1]. Whether or not there is here any "paradox" [1; 8], they should be related to a much older "paradox" inherent in the probability theory itself since the early days of Pascal, Fermat, Laplace and Bayes where it was termed the "principle of probability of causes" [5; 6; 21]. This has been since related to the empirical fact that blind statistical *prediction* is "physical" while blind statistical *retrodiction* is not (i.e. one can calculate the probability that something physical *will* happen but not the probability that something *did* happen). This is recognized today as an asymmetric initial (*a priori*) condition *imposed* by the statistical physicist upon (symmetric) macroscopic evolution equations. According to Beauregard [5; 6; 21], this a *priori* "boundary" condition (which is an *initial* not a *final* condition), reads "*blind retrodiction forbidden*"—very much as the boundary condition in macroscopic wave theories reads "*advanced waves forbidden*", two wordings for one and the same statement. Thus, according to Beauregard, there is a close connection between the principle of wave retardation, the temporal application of Bayes' probability postulate and the causality concept (see Fig. 1). The Einstein—Ritz controversy (in which Einstein maintained that the law of wave retardation should follow from the principle of probability increase while Ritz insisted on deducing the law of entropy increase from the principle of wave retardation) is thus resolved. Some other attempts to account for the origin of irreversibility in statistical mechanics were made.[1]) Yet they also ignore the basic time asymmetry embedded *a priori* in the theory. At this point it might be interesting to note that very significantly Hoyle and Narlikar, and Narlikar [6; 16] have recently demonstrated how the expanding space generates the electromagnetic arrow of time out of the time-symmetrical equations of electrodynamics (see also Fig. 1).

In closing this section I stress only four points: 1. According to Mehlberg [27] Watanabe [33], Reichenbach [31], Adams [20], Grünbaum [24], and others the mathematical expressions of irreversibility possess a "*fact-like*" character. 2. The *introduction of statistics does not, by itself, produce irreversibility* [10]. *One must add a specific assumption of asymmetry* (*an initial condition*) *to the statistics* (i.e. irreversibility and time asymmetry in statistical mechanics are *not* to be found in the mathematics but rather in the *deliberate choice* by the statistical physicist of a *de facto* operational *procedure* which is recognized to day as an imposed initial condition—*essentially equivalent to an a priori introduction of a "second law"*). 3. The (classical or quantal) statistical school *deduces* the dynamics and thermodynamics of the universe *as a whole*, or of *any finite region of it*, from relaxation times of small *local* systems. *This leads to contradictions with observations* [3]. This is a second failure of the statistical school which is related to our discussion. However, it is restricted only to large-scale astrophysical processes. 4. Essentially the statistical and astrophysical schools agree [27] that macroscopic

[1]) Some other authors look for the origin of irreversibility in the nature of the Hamiltonian and the interaction terms which occur in it (cf. e.g. [17]), in the coarse graining of phase space which is required to take account of the fact that all measurements are macroscopic (cf. e.g. [35]), in the passage to the limit of an infinite number of degrees of freedom (cf. e.g. [41]), in the interpretation of the Liouville equation (cf. e.g. [42]), or as due to the impossibility of completely isolating a system from the rest of the universe (cf. e.g. [46]). It should be stressed, however, that while the foundation of AST is not based on Mach's principle, neither is it in disagreement with it. Neither it postulates the existence of an "isolated system".

irreversibility is extracted from time-symmetrical equations *via* some sort of an initial condition. The disagreement is, therefore, inherently about the nature and origin of "proper" initial conditions.

III. The Astrophysical Revolutionary School of the Origin of Irreversibility

Left without any other convincing arguments as to the physical mechanisms and links which originate irreversibility and time asymmetries in the local or large scales of the world, many authors [5; 6; 7; 10; 11; 12; 15; 19; 24; 26; 29; 34; 39; 40] tend now to accept our thesis that irreversibility is essentially cosmological in origin. However, little agreement has been reached so far, about the precise nature of the physical links between the "large" and the "small" systems.

In this section I shall very briefly discuss the questions raised in the introduction, the philosophy of the main theory, its inherent difficulties and the main criticism which it encounters. For more details the reader is referred to references [2; 3; 5 and 7].

1. Entropy vs. the Third Cosmological Time Anisotropy

At the very core of thermodynamics one finds the distinction between reversible and irreversible processes which is related to the "second law" of thermodynamics. Statistical and other formulations of the "second law" require that the "*entropy of any isolated macroscopic system will never decrease*". Therefore, on must merely look at an "isolated" macroscopic system at two instants of time, t_1 and t_2, and determine at which the entropy is greater. That must be the later instant pointing out which way time is advancing in macroscopic physical systems (i.e. whether $t_2 > t_1$ or $t_2 < t_1$). One can then define the "*arrow of time*". In this way thermodynamics shows that the world is unsymmetrical with respect to past and future. However, it is a remarkable fact that the laws of dynamics and microphysics (except the case of the decay of neutral kaons [12; 28]) are symmetrical with respect to time reversal. That is, if one mentally reverses actual motions of particles in a macroscopic physical system (i.e. by passing them through the same sequence of configurations in the reverse order), one gets a process which satisfies the laws of dynamics but may rarely, if ever, take place. This irreversibility appears also when a dipole aerial in empty space generates electromagnetic waves radiating outwards to infinity. However, the time inverse of these processes never happens. For systems composed of large number of particles it is known as the *Loschmidt's paradox*. Much has been written on this subject (see for example references [11; 14]) and there is no need to repeat here the various evaluations of this "paradox".

Referring now to questions 2 and 3 in the introduction and attempting to apply statistical (or classical) thermodynamics to the entire universe, regarded as an isotropic system, a glaring contradiction between theory and experiment arises. According to *statistical thermodynamics*, the universe, or any finite region of it, however large, *should have a finite relaxation* time, and should be in equilibrium. Observations, however, contradict this "prediction". The escape from this contradiction in the statistical school, according to Landau and Lifschitz [13], is to be sought in general relativistic cosmology. When statistical properties of bodies are considered, they maintain, the metric properties of space-time may be regarded as non-steady "external conditions" since the metric tensor is, in general, a function not only of the space co-ordinates but also of time.[2]) Thus the universe as a whole, or any finite region of it, must be considered as a *non-equilibrium system* in a variable gravitational field. Accordingly, the classical formulations of the "law of increasing entropy" are not valid for larger and larger portions of the universe. The description of largescale phenomena in space poses another inconsistency to the assumed *additivity of entropy*. Up to a given volume of space the quantity of *information* required to describe the contents of the system undoubtedly increases in direct proportion to the volume. However, the contents of a volume whose dimensions greatly exceed the scale of local irregularities is largely predictable (i.e. the accuracy of predictions in cosmology largely increases with volume [26]). Since only a finite quantity of information is required to specify the entire universe,[3]) the *entropy per unit volume tends to zero as the volume increases indefinitely*. Therefore, *the very concept of additive entropy fails*. These contradictions are intimately coupled with the well-known *Olbers*' "*paradox*" according to which the night sky should be intensely brilliant in a *static* universe [5; 7; 11; 16]. However, the solution to the last "paradox" was found when it was discovered that the universe expands and radiation, therefore, flows into cold, nonreflecting space (see Fig. 1). *Empty space acts therefore as a low-temperature thermal reservoir* and one might say that "entropy increases" *results* from "temperature equalization" with this cold "*universal sink*". This is the most striking fact of the third *cosmological arrow of time* (see below and Fig. 1).

2. The Three Cosmological Time Anisotropies

In this section I distinguish among at least two cosmological arrows of time. The first one, and according to the astrophysical school [5; 6; 7; 15] the *Master* one, is the time asymmetry caused by the expansion of space. Thus, if we photograph a group of supergalaxies at instants t_1 and t_2, their different separations from each other would tell us whether $t_2 > t_1$ or $t_2 < t_1$. Hereafter we refer to this time asymmetry as the *first or the Master cosmological arrow of time*. According to Gold [5] and Gal-Or [7] *it generates and dominates all other time asymmetries that are in evidence*. I shall return to this point later. This Master arrow is, of course, independent of any particular model of expanding (or contracting) world. For *time-dependent* cosmo-

[2]) Time, on the other hand, is fundamentally connected with the expansion of the universe as demonstrated by equation (6) below.

[3]) Which, in accord with all our observations, is isotropic.

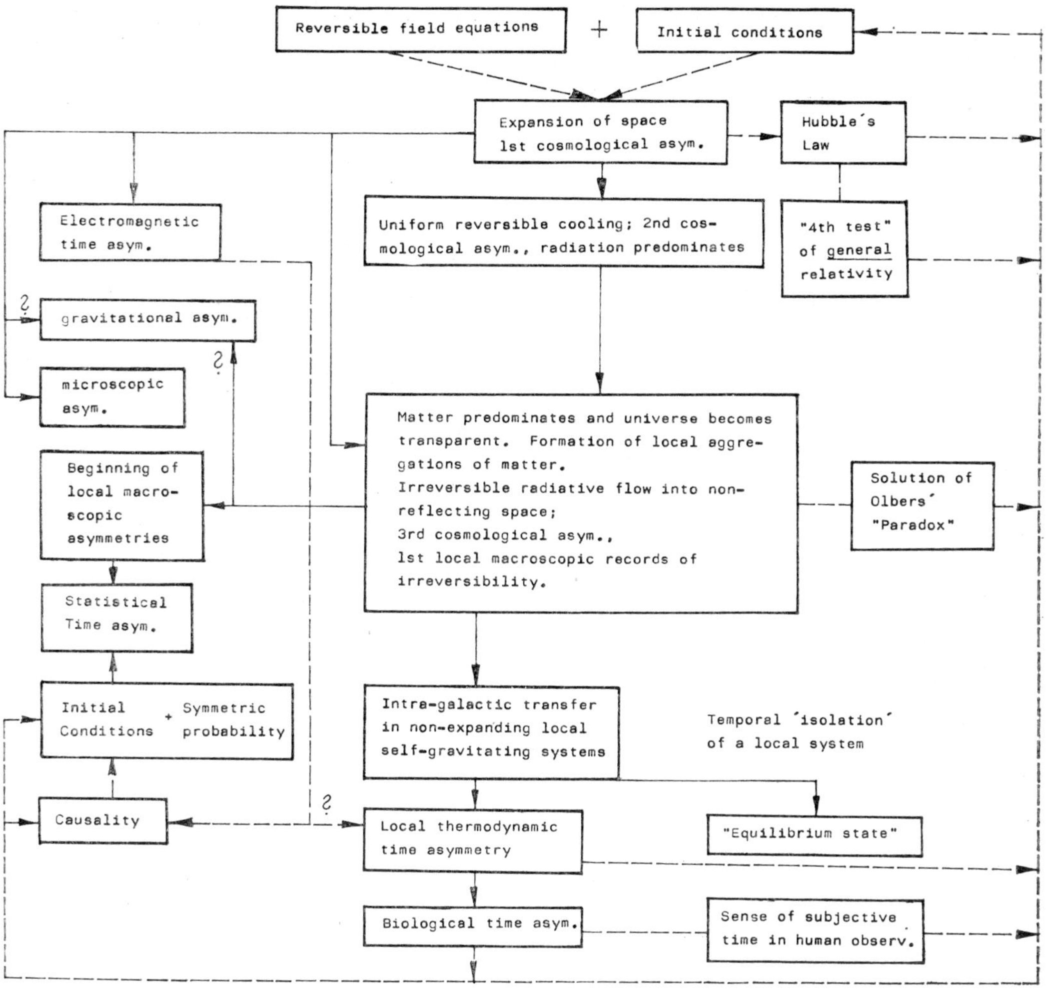

Fig. 1. The main, one-to-one links, according to the astrophysical school of the origin of irreversibility in nature [2]. The one-sided dynamics of the world expansion (the master arrow) *originates* all other time arrows, and causes them to be consistent with each other. *Mutatis mutandis*, the conclusions of the astrophysical school are in agreement with all observations now in evidence on the local as well as cosmological scales. It is shown that none of the other physical asymmetries can generate the master arrow (only the reverse links become possible). Yet, this school leaves the philosophical question of the proper choice of cosmological initial conditions largely open. (Solid lines represent assumed physical links.)

logical models, I define a *second cosmological arrow of time* by the decreasing temperature, or density of radiation, in an expanding world. For instance, the temperature T of the black body radiation may be related to the linear scale factor $R(t)$ of the universe as [45]

$$T \propto R^{-1}(t). \tag{1}$$

A similar arrow[4]) may be defined by means of the energy density of the radiation field ϱ_{rad} which obeys the relation

$$\varrho_{rad} \propto R^{-4}(t). \tag{2}$$

By contrast, the density of matter (if it is conserved) obeys the relation

$$\varrho_{mat} \propto R^{-3}(t). \tag{3}$$

According to "hot big bang" cosmological models, $R(t)$ was arbitrarily small in the past, and so, if there was any radiation at all, its energy density dominated that of matter at sufficiently early times (see Fig. 1).

A radiation dominated universe is easy to handle in general relativistic cosmology with the simple results

$$R(t) \propto t^{1/2}, \tag{4}$$

$$T_{rad} \propto \frac{10}{t^{1/2}} \tag{5}$$

for small values of t (t is expressed in seconds). At much later times two important things occur: As the expansion proceeded the temperature of the *uniform* medium fell, till, at a about 3000 °K electrons were captured by nuclei and remained in bound orbits. At this stage they no longer scattered photons and the *universe became transparent*. Radiation then ceased to be strongly coupled to matter and it also ceased to dominate energetically [45]. This second cosmological time asymmetry is, of course, *not* accounted for in the steady-state models. However, the evidence against the steady-state theory is now quite strong. But in its time it played an important role in forcing its developers to devise their theory of the origin of the elements in hot stars [45].

I now define a *third cosmological time arrow* which is accounted for in *all* cosmological models, as well as observations which are now in evidence. It is the one-sided *flow of radiation* energy *from* the surface of all galactic systems *to* the cold, non-reflecting expanding space which soaks up any amount of radiation released in the stars. *The second and third arrows are, therefore, originated and dominated by the first one* (*while the reverse links are physically impossible*).

3. The Universal Thermodynamic Sink

The observed dark sky and red-shifts prove that thermodynamic equilibrium does not prevail throughout the vast regions of the expanding space. Therefore, the expanding space becomes a huge "*thermodynamic sink*" for all the radiation flowing *out* into empty space. This conclusion is not only supported by cosmological observations but it agrees with the theoretical predictions of general relativistic cosmology. Perhaps no test of Einstein's theory of relativity is more

[4]) We are concerned only with the asymmetry and not with the values of change.

dramatic than the expansion of the universe itself.[5]) If Einstein had held fast to his original theory (without introducing into it the so-called cosmological constant to avoid Friedman's prediction of an expanding universe), Hubble's discovery, years later, could hardly have been considered as any but the fourth test of relativity. Very significantly, and in analogy to the statistical school, this basic cosmological arrow of time can be *theoretically deduced from time-symmetrical theories of gravitation, including Newtonian dynamics and the field equations of relativity*. We therefore conclude that fundamentally thermodynamics and gravitation are connected. (For a proof see [3].)

4. The Dominance of the Master Arrow

At this point one may wonder *how* the Master cosmological arrow may dictate, according to the astrophysical school [5–7; 15], *all* processes far away along the world lines? What possible meaning does it have to talk about the expansion of the universe and its effects on *local* macroscopic processes, say on earth? Why should the *large* system dictate the behaviour of the *small* system, and not *vice versa* as is usually believed?

Before trying to answer these questions one must note first that the actual expansion is *not* homogeneous throughout space. The radius of the Earth's orbit, for example, does *not* change during the expansion. The size of our galaxy does *not* increase either! What increases is the *separation* from one supergalaxy to another. Yet, these facts are consistent with other observations, which demonstrate that the visible part of the expanding universe is, as far as we can tell, roughly *homogeneous and isotropic*. According to Gold [5] and Gal-Or [7] the radiation expansion going on around most galactic material (the third cosmological arrow) gives rise to *all* other local time asymmetries and irreversibilities that are in evidence in the stars and on earth (Fig. 1). To demonstrate this point Gold [5] points out that all time's arrows eventually will be lost if a star on which they appear can be "completely isolated" in an "imaginary box" which would prevent the expansion of radiation out into space.[6]) This general conclusion agrees with Narlikar's [6; 16] derivations which deduce the electromagnetic time asymmetry from the expanding universe. In electrodynamics, one must note that an accelerated electric charge *loses* energy (retarded solutions). Thus the energy flow from the charge can be used to fix the direction of time. However, the proper original equations must be time symmetric. If advanced solutions are chosen the *charge receives* energy. Why, therefore, do we choose retarded solutions? The common answer is "*because of causality*"! However, to say that one must use retarded solutions in order to conform with causality does not advance us any further in our understanding of the *origin* of the electrodynamic time asymmetry.

[5]) The other three tests are, the precission of the perihelion of Mercury around the sun, the bending of light by the sun, and the redshift of spectral lines by the gravitational field of the sun.

[6]) Gold's "star in an imaginary box" should be replaced with "a supergalaxy in an imaginary expanding cell". For details see [3].

A progress in this direction has been achieved by Hogarth [1] and Narlikar [6; 16], who improved the arguments employed in the Wheeler-Feynman theory for a *static* universe, by considering the first (Master) time asymmetry as a generator of retarded solutions.

5. *Intra-Galactic Transfer and Local Thermodynamics*

The main difficulty in explaining local irreversibility is to show *how* the thermodynamic (and electromagnetic) time asymmetries are linked to the first and the third cosmological arrows in a *non-uniform expanding world.* What might be the mechanism in *non-expanding* galactic systems such as on earth? In trying to answer this questions Gal-Or [3; 7] maintains that the mechanism of transfer of energy in a non-uniform expanding universe must be thought of as being sub-divided into two different mechanisms. The *first* transfer of energy is carried out, as previously described, by the radiation expansion from the surface of hot galactic bodies, into the nonreflecting, expanding, inter-galactic space which thus becomes a huge "thermodynamic sink". The *second* transfer mechanism occurs inside *non-expanding* galactic bodies, and involves a very large number of *intermediate* stages and intermediate reservoires. As we know from earth bound observations this transfer mechanism is much more complicated since in addition to radiation, energy is released and transferred by such local processes as conduction, convection and chemical and nuclear reactions (see Fig. 2). These processes are briefly described in the next two sections.

a) Stellar Energy Losses and the Astrophysical Thermodynamics

In the stars nuclear and gravitational energy is transformed into photons and neutrinos. The star lives on its nuclear resources during a number of different nuclear burning stages (see Fig. 3 and 4). During such stages the *rate of energy release* from any particular nuclear reaction *is just sufficient to compensate for the radiation energy loss into cold outer space* [44]. The star stops its gravitational contraction for the time being, to resume it only after exhaustion the particular fuel involved. Fig. 3 and 4 show that in the *interiors* of highly evolved stars

→ Fig. 2. *Intermediate Cycle of Solar Energy in the Earth's Atmosphere, Before it is "Lost" in Non Reflecting Expanding Space. All Thermodynamic Processes are Caused by these Input-Output of Energy Fluxes.* If one postulates a sudden hypothetical equilibrium of the upper atmosphere with the radiation in outer space, all input-output energy fluxes will stop, and all irreversible processes on earth will decay within a finite relaxation time.

In this sense, the very existence of the cold, non-reflecting "*universal sink*" is the "cause" for the generation of gradients and time asymmetries in the earth's atmosphere. Thus, the master asymmetry, indirectly, generates and dictates thermodynamic cycles on earth or in other stellar atmospheres. (Numbers in the figure are based on an assumed solar constant of 1395 Wm^{-2} and a mean global albedo value of 0.34 ε denotes an average rate of less than 0.5 Wm^{-2}. Note that, if, for the sake of illustration, all energy inputs for the atmosphere cease, and rates of energy expenditure are maintained, the reservoir of internal energy (thermal) will be depleted in about 100 days; the reservoir of latent heat (precipitable water) in about 12 days; and the reservoir of mechanical energy (momentum) in about 3 days. Figure modified from reference [38].

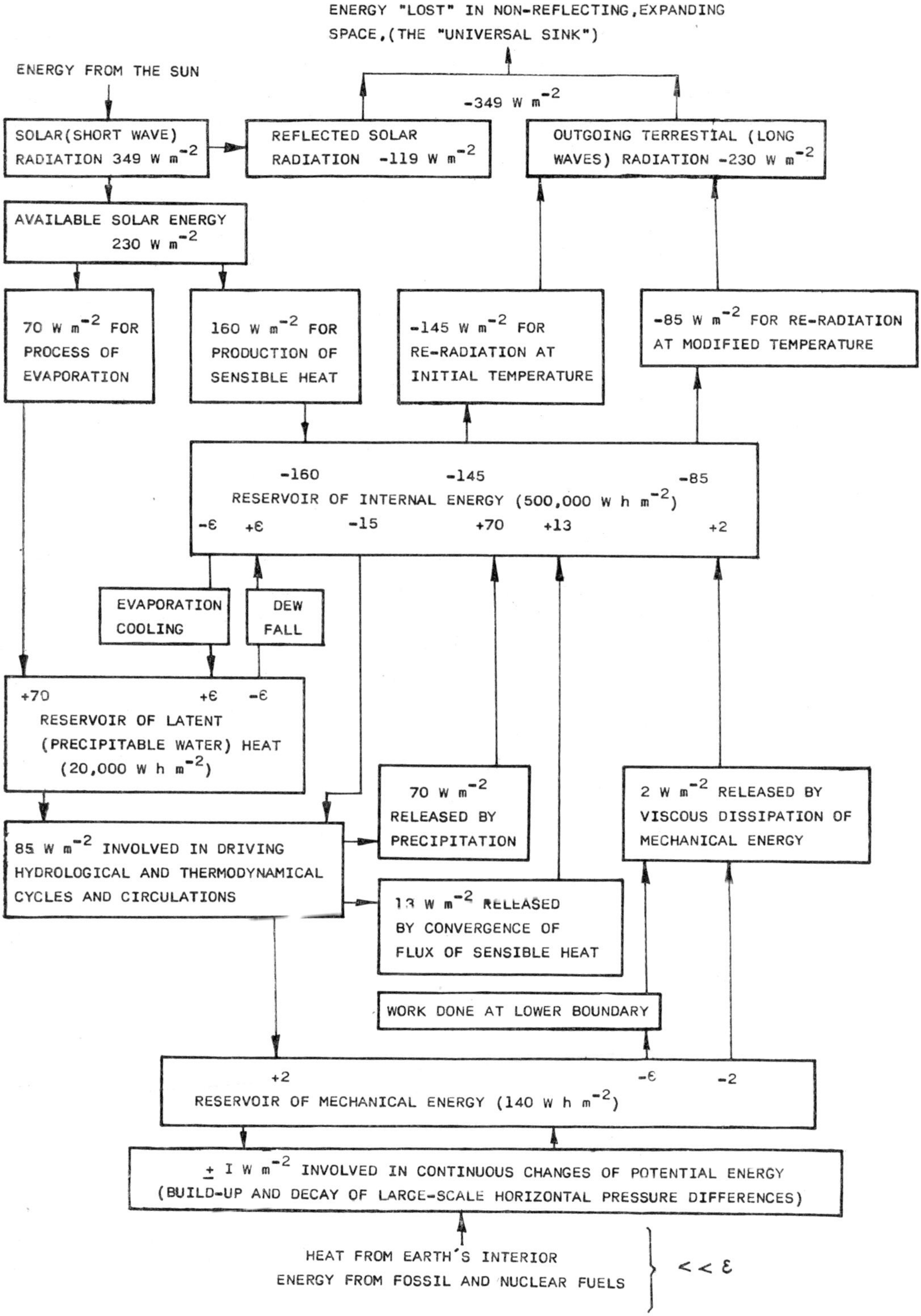
ENERGY "LOST" IN NON-REFLECTING, EXPANDING SPACE, (THE "UNIVERSAL SINK")
ENERGY FROM THE SUN
-349 W m^{-2}
SOLAR (SHORT WAVE) RADIATION 349 W m^{-2}
REFLECTED SOLAR RADIATION -119 W m^{-2}
OUTGOING TERRESTIAL (LONG WAVES) RADIATION -230 W m^{-2}
AVAILABLE SOLAR ENERGY 230 W m^{-2}
70 W m^{-2} FOR PROCESS OF EVAPORATION
160 W m^{-2} FOR PRODUCTION OF SENSIBLE HEAT
-145 W m^{-2} FOR RE-RADIATION AT INITIAL TEMPERATURE
-85 W m^{-2} FOR RE-RADIATION AT MODIFIED TEMPERATURE
-160
-145
-85
RESERVOIR OF INTERNAL ENERGY (500,000 W h m^{-2})
-ε
+ε
-15
+70
+13
+2
EVAPORATION COOLING
DEW FALL
+70
+ε
-ε
RESERVOIR OF LATENT (PRECIPITABLE WATER) HEAT (20,000 W h m^{-2})
70 W m^{-2} RELEASED BY PRECIPITATION
2 W m^{-2} RELEASED BY VISCOUS DISSIPATION OF MECHANICAL ENERGY
85 W m^{-2} INVOLVED IN DRIVING HYDROLOGICAL AND THERMODYNAMICAL CYCLES AND CIRCULATIONS
13 W m^{-2} RELEASED BY CONVERGENCE OF FLUX OF SENSIBLE HEAT
WORK DONE AT LOWER BOUNDARY
+2
-ε
-2
RESERVOIR OF MECHANICAL ENERGY (140 W h m^{-2})
± 1 W m^{-2} INVOLVED IN CONTINUOUS CHANGES OF POTENTIAL ENERGY (BUILD-UP AND DECAY OF LARGE-SCALE HORIZONTAL PRESSURE DIFFERENCES)
HEAT FROM EARTH'S INTERIOR
ENERGY FROM FOSSIL AND NUCLEAR FUELS
<< ε

the highest temperature is perhaps 10^{10} °K, which is high enough to convert matter to the most stable element, iron. It is the energy *generated* by these processes which according to the astrophysical school, must be transported by radiation, conduction, convection or by mechanical means (such as expansion or pulsation) *from the interior of the stars to their surfaces, where it is radiated away to the cold, non-reflecting "universal sink". These transport processes*

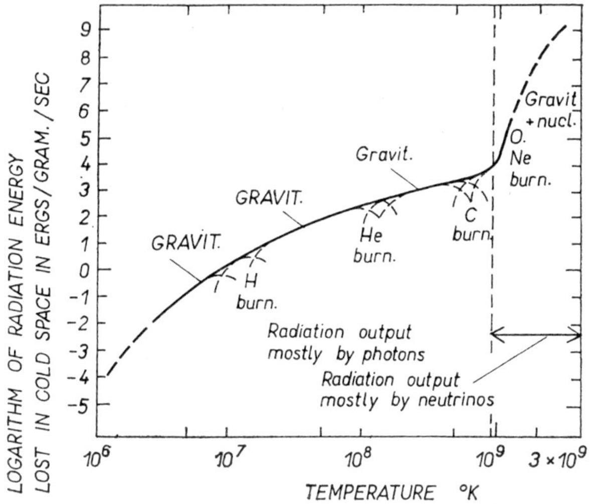

Fig. 3. A rough sketch of the over-all energy *generated* by a star (of about one solar mass), which is *transformed into radiation "lost" in cold, nonreflecting space as photons and neutrinos*. The origin of the radiated energy is identified by "Gravit" for energy released by gravitational contraction and by the chemical symbol of the nuclear fuel for the radiation losses which originate from the nuclear burning stages. Dashed lines represent individual contributions of these two modes while the solid line is their sum (the total luminosity of the star in erg/gm/sec). The region where luminosity switches from mostly photons to mostly neutrinos is indicated. (After Reeves [44].) For more information see Fig. 4.

depend, therefore, and as a matter of fact are driven by the very existence of the cold, non-reflecting "universal sink" (or by its origin – "the expansion of space"). Temperature gradients, from the interiors of the stars to their surface, can be approximated by about 10^7 °K over a radius of 10^8 km in a typical star [43]. *The formation of temperature gradients is a requirement in stellar interiors. It is produced primarily by the condition of hydrostatic equilibrium, which causes the energy to flow outward to the surface and from there to the "universal sink".* Clearly in a static ("equilibrium") world these gradients and irreversible processes would have become impossible.

b) Earth's Energy Losses and the Local Thermodynamics

If one postulates the establishment of a sudden hypothetical equilibrium of the earth's upper atmosphere with the radiation density in the outer space then *all the input-output energy fluxes* to earth *should stop* and all irreversible processes on earth *will be decaying* within *a finite relaxation time*. Fig. 2 demonstrates that when all energy inputs for the atmosphere cease and rates of energy expenditure are maintained, the *reservoir of internal energy* (*thermal*) *will be depleted in about 100 days; the reservoir of latent heat* (*precipitable water*) *in about 12 days* and the *reservoir of mechanical energy (momentum) in about 3 days!* I therefore conclude that *all the intermediate thermodynamic cycles of solar energy* in our pseudo-steady-state atmosphere,

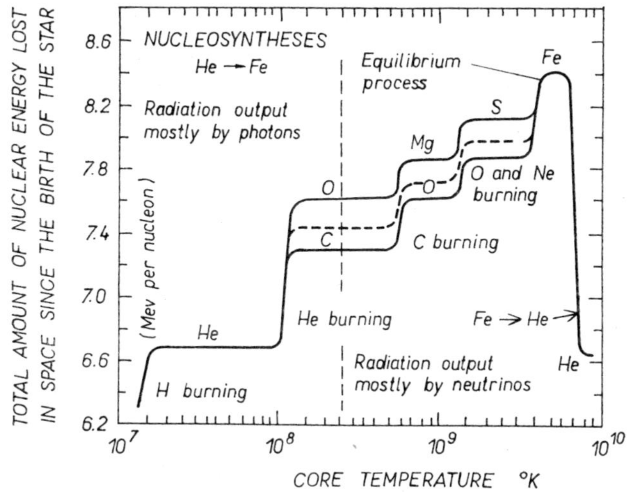

Fig. 4. *Total amount of radiation energy "lost" in cold, non-reflecting space from nuclear burning stages* (per nucleon). During the nuclear burning stages, the rate of energy *released* in the *core* of the star *transforms* into an *equivalent* rate of *radiation loss* into cold, non-reflecting expanding space (the "*universal sink*"). Gravitational contraction (in this diagram represented by the flat regions) is resumed after exhaustion of the particular fuel involved. Note the *overall reversible* chemical cycle He → Fe + Energy, which infers the possible formation of an equilibrium state (analogous to chemical reactions in closed systems) and *loss of irreversibility* if one postulates "*hypothetical equilibration*" with *radiation density in outer space* (i.e. a *static* universe). (After the He-burning stage, the core contains C^{12} and O^{12} in fractional abundances depending upon the initial mass.) The C and O curves are, then, lower and upper limits, respectively, as they represent cases in which the core is pure C or pure O. To the left of the vertical dashed line, most of the energy is lost in space as photons; to the right, as neutrinos. The dashed line is a typical case. The same kind of statement applies to the other nuclear burning stages. After Reeves [44].)

are intimately *coupled* (through large, but finite relaxation times) *to the mechanism by which solar energy is eventually* "lost" by *re-radiation* into cold, non-reflecting, expanding outer space. In this sense the *very existence* of the non-reflecting, cold, "universal sink" is the *cause* for the creation of *all* gradients in the earth's atmosphere (or, as we have seen, in any other star).

Yet, one may ask how do we account for the existence of "*equilibrium states*" in smaller, *local* systems on earth? First, let me stress that *relaxation times increase with the size of the system*. Therefore, separate *small* parts of the system attain "partial" equlibrium states considerably *more quickly* than bigger systems do. If the entire earth's atmosphere is not in an equilibrium state but in a pseudo-steady state, small *local* regions can still attain "partial" equilibrium states *while the whole system remains in a non-equilibrium state* (e.g. the change of temperature with altitude). Now, the *over-all* heat balance of the earth's pseudo-steady-state atmosphere, *requires*, that *any* artificially produced energy *source* in the atmosphere, or on the ground (say, by combustion of fossil fuels or use of nuclear energy), should either cause an *additional* flow of energy (via many intermediate stages), *outward* into cold non-reflecting space (by radiation transfer), or (especially if the sources' intensities gradually increase with time), *change the existing pseudo-steady-state into new ones. One must radically reject any trace of traditional notions that the earth atmosphere, or its oceans, are "infinite heat resorvoirs"*. The now fashionable "world pollution crisis" and its links to the earth's albedo, temperature, etc. is only a minor awareness or manifestation to this global mechanism. I therefore conclude, by stating, that the time constant for any intra-galactic transfer depends upon the *scale* and *structure* of the *non-expanding* galactic system. *In the laboratory*, on the other hand, the time constant for such an intra-galactic transfer may be estimated by a *temporal* 'isolation' of a macroscopic system, during which irreversibility persists for a while but decays from a high value initially, to zero eventually. When the time-invariant "state" is reached the system is said to be in a "*state*" of "*equilibrium*". This *local* "equilibrium state" may, at the laboratory scale, be extended in time by *imposition* of "time-invariant" boundary conditions via an interaction with *other* local systems which supply the required energy (and are, therefore, *not* in an "equilibrium state", since they form part of the rest of the universe, which is never in an "equilibrium state"). This mechanism thus expalins the occurrence of local "equilibrium states" in an *over-all non-equilibrium surroundings*. Practical (local) thermodynamics is frequently concerned, however, with the relaxation times which bring the local system to *such* an equilibrium state. Yet, so-far, it misjudged the *origin* of irreversibility and entropic growth in these local processes. Irreversibility is enclosed with the medium at the moment of "isolation". Being cut off from the external world asymmetries its enclosed irreversibility decays. It can never reappear unless reconnected with the outer world for a fresh influx of irreversibility. Its origin is, therefore, never to be found inside the local system being considered.

6. Origin of Thermodynamics and the Expanding Box

In supporting our school of thought Beauregard maintains that if a *reflecting box is not expanding*, all enclosed quantal systems will keep fluctuating around an "equilibrium energy state", and exchanging, quanta on "stationary" waves. However, if the reflecting box *is expanding*, all enclosed quantal systems will be *decaying* and *emitting* quanta on all sorts of electromagnetic, electronic, nutronic, etc., waves. More specifically, Beauregard's feels that space expansion goes hand in hand with a general decay of gravitating sub-systems and general *emission* of *gravitons* on the corresponding retarded waves.

In general, I maintain that if the reflecting box (which represents a cluster of galaxies in *isotropic,* expanding world) contains self-generating energy sources, it would never reach thermodynamic equilibrium as long as the box is maintained in a state of sufficiently fast expansion.

7. Concluding Remarks

To summarize this section I specifically stress the following concluding remarks: *None* of the physical links shown in Fig. 1 *can generate the world expansion. Only the reverse links become possible.* Consequently, the astrophysical school claims that it is impossible for the *small* system to *dictate* irreversibility in the *very large system* (as implied in current theories), *but only the reverse sequence of links is possible.* We must, therefore, abandon the foundations of local thermodynamics and call the large scale dynamics of the universe to our aid in order to build a new theory for describing, *both,* the local and cosmological phenomena *on the same footing.* Another conclusion which is intimately linked to the new concepts of the astrophysical school is related to the origin of the *microscopic* time asymmetry in *elementary particles.* Here, Aharony and Ne'eman [12] have recently demonstrated (for the evolution of the $K^\circ - \bar{K}^\circ$ (kaon) complex) a possible relation with the cosmological asymmetry. When the Master arrow is thus taken as the most fundamental time asymmetry, the formulation of a unified thermodynamic theory may be started directly from the proper field equations which govern the very expansion of the universe as a whole (most likely those of general relativity). *The "second law" of thermodynamics is, therefore, not an independent postulate or "law".* It is based on initial cosmological conditions and must be deduced and derived (in terms of energy units) from the local conservation of energy (the first law of thermodynamics), on one hand, and the global dynamics of the evolving world on the other hand. Adopting this approach it would seem possible to reduce the *total* number of axioms and postulates—in particular the existence of a unique entropy value in a state which is attained during an irreversible process (cf., e.g., [1, pp. 37, 47]), the additive properties of entropy, the "fact-like" axion of the "second law" (e.g. the Carnot-Clausius-Plank inequality), the existence of "isolated" systems, the "third law", and the LTE assumptions involved with the very concept and use of temperature. The immediate aim must therefore be the construction of global-local thermodynamic theory without an *a priori* assumption of the existence of entropy and with a redefined temperature function. In this way the new Astrophysical School of Thermodynamics takes the *observed* expansion as the point of departure in laying new foundations for thermodynamics [7; 27]. This theory is first of all based on observations and facts in the local as well as global frames of reference. Secondly, it conforms with independent *theoretical* predictions of general relativistic cosmology as well as with the rest of the conclusions of local thermodynamics. Third, it avoids a number of paradoxes, inconsistencies, and misconceptions inherent to current thermodynamic theories. Therefore, the main difficulty which challenges this school in the near future, is to formulate proper links between the world of the very large and the world of small, local, macroscopic systems *within the frame of a single, entropy-free, unified theory of thermodynamics* (see Note Added in Proof).

IV. Criticism of the Astrophysical School

Criticism of the astrophysical school comes from the classical and statistical schools. Very convincingly the Copenhagenian school [9] maintains that *irreversibility originates in the macroscopic instrument* (or observer) which records information and which retains a record of it (but which did not have the record in the past). Accordingly, the asymmetry in time may be *imposed by the observer* and *not* created by the system of which the observer forms part. In short, *the choice of initial conditions is not a law of nature but only a result of our position as macroscopic observers*. The defenders of the astrophysical school usually respond to this criticism by noting that we, as human observers, cannot *cause* macroscopic systems (much less the whole universe) to run *backwards* by our personal choice or very presence. Furthermore, observers cannot be isolated from the rest of the universe and are, therefore, not entirely independent. Accordingly, initial conditions may have the character of laws, or "auxiliary conditions [26], *on the same epistemological basis* as any of the other reversible laws of nature which are intimately coupled to observations. Yet, observations are limited to what our epistemological position allows, (e.g. we cannot distinguish between two possible symmetric universes – of matter and antimatter – since we happen to live in only one of them [5]). We calculate only the probability that something *will* happen but not the probability that something *did* happen since we live in *one-sided* expanding universe which *had originated by an initial and not a final condition*. In this way the *one-sidedness* of the dynamics of the universe dictates, *a priori*, the irreversible nature of observations which generate thermodynamics. In trying to reconcile the two schools of thought I raise the question; what might be the *earliest possible local macroscopic instrument* to record irreversibility in an expanding (or contracting) universe? My lemma is that the origin of a *so-defined* irreversibility can be traced backward in time *maximum* to the period when the first local *self-gravitating aggregations* of matter *had started to commence in the expanding universe*. This proposition is based on the well-accepted assumption that *before* the formation of local irregularities of matter the universe was *uniform*, with radiation making an overwhelmingly dominant contribution to the pressure (in equilibrium with uniformly scattered nuclei). As the expansion proceeded, the uniform temperature fell reversibly (the second cosmological asymmetry), till, at about 3 000 °K, electrons were captured by nuclei and remained in bound orbits. At this state they no longer scattered photons, and the universe became transparent. Therefore, aside from the problem that we are still far from understanding in details the *formation* of self-gravitating aggregations of matter, one arrives at the conclusion that local macroscopic information of irreversibility *was not in existence at the earliest period* (at least 100 000 years) while there was *already* at least one clear and dominating cosmological time asymmetry. Consequently, to affirm one school and deny the other would result in a contradition in a *non-uniform* expanding world; however, in considering a *uniform* expanding world, the astrophysical school supplies a much more universal theory about the *original asymmetry of nature* and the *sequence* of relationships which *link* it to all asymmetries that *are now in evidence*. The problem of man in the small, and the universe in the large, are thus *not entirely disconnected* from each other. *For life to exist, with its own biological arrow of time, there must exist temperature gradients*, *which in turn had been originated in the expanding universe*. Therefore, there may be *some links* between initial conditions, expansion of outer space, thermodynamics, and the *origin of life* (see Fig. 1). In other words, it is reasonable to assume, along this pattern of thought that life (as well as all other time asymmetries), could *not* have been developed in a *static* universe. While the answer to the *last point may never* be known with certainty it is more reasonable to think that our *subjective* understanding is inadequate and is perhaps preventing us from a better description of the origin of the time concept. No compulsive or conclusive answer should be imposed now on these fundamental questions. They should be kept open, till, with the passage of time, some new physical theory may be devised that would define entirely new concepts of time and time asymmetries. Perhaps we ought to hold to a somewhat skeptic attitude that, in Sir James Jeans's words, to many it is not knowledge, but the quest for knowledge, that gives the greatest interest to thought—to travel hopefully is better than to arrive.

Is that the end of the road in thermodynamics? Apparently, the answer is no! There may be some guiding clues in the new concepts of superspace [10; 34] which are related to initial phases of the universal expansion [2], in new experiments with kaonic systems, in new concepts about the role of neutrinos in expanding space, and in more refined mathematical analyses of Maxwell equations in expanding world.

V. Absolute (Anisotropic) Time

It is highly doubtful if rates of cosmological processes, and the age of the universe, should be measured by any of the present standard units of time. During the early stages of the Universe, which involved extremely high temperatures ($> 10^{16}$ K°), all units, such as earth years, atomic clocks, nuclear transitions or muon lifetimes, were not relevant or not in existence [18]. Yet, the dominant *Master time asymmetry* was there since the beginning of the expansion. Why not to use, therefore, this master (volumetric) asymmetry *to redefine the very concept of time?* For instance, I find it logical as well as useful to define time by length according to the simplest link

$$t = V^{1/3} \tag{6}$$

where V is the volume of space at a given epoch. The choice of volume instead of a linear scale dimension is intended to avoid possible anisotropy in the early phase of the cosmic expansion [18]. *The evolving universe itself must therefore, be considered as an absolute clock.* Eq. (6), therefore, defines an *anisotropic time*, which is always subject to the condition

$$t \geqq 0. \tag{7}$$

This simple definition is *not only relevant during the entire lifetime of the expanding universe*, but it may also appeal to us for a *contracting* space. This concept becomes useful in particular for the *entropy-free thermodynamics* [22], since even in a contracting space

$$\frac{dV}{dt} > 0 \tag{8}$$

and our formulations remain valid. Eqs. (6) and (8) (or any other definition of time) may pose interesting questions when applied to an isotropically *oscillating* universe. *Suppose* an "observer" managed to survive the transition through the epoch of maximum expansion (perhaps, by shutting himself in a shelter), so that his own time asymmetry remained unchanged. Emerging later in the contracting phase, his time asymmetry would contradict asymmetry (6) (or much later the time asymmetry due to the 3rd cosmological asymmetry) but would be in agreement with (8), especially if (8) *is* to be taken as the *Master asymmetry* (with initial and final conditions which are correspondingly connected to expanding and contracting phases of oscillating cosmologies [40]). Due to the *large time constants* involved, the *reversal* in the *direction of energy flow* (the 3rd cosmological asymmetry) may *not* occur at the epoch of maximum expansion. The analysis of the transition stage remains, therefore, still obscure since it *may* involve a period with *opposite asymmetries* (e.g. the 2nd and 3rd cosmological asymmetries). However, *if the approach to maximum expansion is sufficiently slow*, the *transition stage* may become a *pseudo-static* one (i.e. like in a static world with "equilibrium" conditions throughout its extent, where all asymmetries may vanish)[7]) and *all asymmetries would closely follow the master asymmetry given by eq.* (8). According to this master asymmetry the dynamic universe *does not accelerate! Nor does it decelerate.* The disadvantage of this concept of time is, however, inherently related to *our present ignorance* as to how to correlate our time with the precise history of V (or, *which* cosmological model should be adopted as the universal standard). However, this concept of time may bear its own merits through the use of eq. (8) even with our present inability to assign t precise numerical values in our times.

Note Added in Proof. Many of the concepts discussed here have been most recently given a wider quantitative and mathematical interpretations in our new book *Modern Developments in Thermodynamics* [3].

References

[1] A Critical Review of Thermodynamics (E. B. Stuart, B. Gal-Or, and A. J. Brainard, Eds.). Mono Book Corp., Baltimore 1970 (The Proceedings of an International Symposium "A Critical Review of the Foundations of Relativistic and Classical Thermodynamics", Pittsburgh, Pa. April 7—9, 1969).

[2] B. Gal-Or: The Crisis about the Origin of Irreversibility and Time Anisotropy. Science (U.S.) *176* (April 7, 1972), 11—17.

[3] Modern Developments in Thermodynamics (B. Gal-Or, Ed.). J. Wiley, N. Y. and Israel Universities Press, Jerusalem 1974.

[4] The Nature of Time (T. Gold, Ed.). Cornell Univ. Press, 1967.

[5] T. Gold in Recent Developments in General Relativity. Pergamon Press, N. Y. 1962, p. 225.
T. Gold; Am. J. Phys. *30* (1962), 403.
T. Gold in [4], p. 1, 128, 184, 229.

[7]) Including, of course, the biological ones of all "observers".

[6] J. V. Narlikar in [4], p. 25, 28, 62.
J. V. Narlikar; Pure & Appl. Chem. *22* (1970), 543.
J. V. Narlikar and F. Hoyle, Nature *222* (1969), 1040.
[7] B. Gal-Or in [1], p. 1, 73, 445.
B. Gal-Or; Nature *230* (1971), 12.
B. Gal-Or; Science (U.S.) (April 7, 1972).
B. Gal-Or; Nature *234* (1971), 217.
B. Gal-Or; Anal. N. Y. Acad. Sci. *196* (A 6) (March 1972), 305.
B. Gal-Or in [3], p. xi, 429.
[8] O. Costa de Beauregard in [1], p. 463.
[9] L. Rosenfeld in [4], p. 135, 187, 191, 194, 227, 230, 242.
[10] P. G. Bergmann in [4], p. 40, 185, 189, 233, 241.
P. G. Bergmann; Trans. N. Y. Acad. Sci. (Jan. 1971).
P. G. Bergmann in Relativity and Gravitation (C. G. Kuper and A. Peres, Eds.). Gordon and Breach, N. Y. 1971.
[11] O. Costa de Beauregard in [3], p. 73.
[12] A. Aharony in [3], p. 95.
A. Aharony and Y. Ne'eman; Int. Journ. Phys. *3* (1970), 437.
[13] L. D. Landau and E. M. Lifshitz: Statistical Physics. Addison-Wesley Press, Reading, Pa. 1969, p. 29.
[14] I. Prigogine in [1], p. 1, 461.
[15] J. Hogarth: Ph. D. Thesis. London 1953.
J. Hogarth; Proc. Roy. Soc. *A. 267* (1962), 365.
[16] J. V. Narlikar in [3], p. 53.
[17] L. Van Hove; Physica *21* (1955), 517 and *25* (1969), 269.
[18] C. W. Misner; Phys. Rev. *186* (1969), 1328.
[19] P. T. Landsberg; Nature *225* (1969), 1204.
P. T. Landsberg (Ed.); Pure & Appl. Chem. *22* (1970), Nos. 3—4.
[20] E. N. Adams; Phys. Rev. *120* (1960), 675.
[21] O. Costa de Beauregard in Proceedings of the International Congress for Logic, Methodology and the Philosophy of Science (Y. Bar-Hillel, Ed.). North Holland 1964, p. 313.
O. Costa de Beauregard: Le Second Principle de la Science du Temps. Edition du Seuil, Paris 1963.
O. Costa de Beauregard; Pure & Appl. Chem. *22* (1970), 540.
[22] B. Gal-Or; Nature *234* (1971), 217.
[23] H. Feigl and G. Maxwell (Eds.): Current Issues in the Philosophy of Science. Holt, Rinehart and Winston, N. Y. 1961.
H. Feigl and G. Maxwell (Eds.): Scientific Explanation, Space and Time. University of Minnesota Press, Minneapolis 1962.
[24] A. Grünbaum: Philosophical Problems of Space and Time. Knopf, N.Y. 1963.
A. Grünbaum and E. L. Hill; Nature *179* (1957), 1269.
A. Grünbaum in [4], p. 149, and in [3], p. 413.
A. Grünbaum; Philos. Sci. *29* (1962), 146 and *37* (1970), 496—588.
A. Grünbaum; Annals of the New York Acad. of Sci. *138* (1967), 374.
[25] J. E. Hogarth in [4], p. 7.
[26] D. Layzer in [4], p. 111.
D. Layzer in Relativity Theory and Astrophysics (J. Ehlers, Ed.). Am. Math. Soc., Providence, R. I. 1967, p. 237.
D. Layzer; Pure & Appl. Chem. *22* (1970), 457.
[27] H. Mehlberg in Current Issues in the Philosophy of Science (H. Feigl and G. Maxwell, Eds.). Holt, Rinehart and Winston, N. Y. 1961, p. 105.
[28] Y. Ne'eman; Proc. of the Israel Acad. of Sciences and Humanities *3* (1969), No. 13.

[29] O. Penrose and I. C. Percival; Proc. of the Phys. Soc. *79* (1962), part 3, No. 509, p. 605.
[30] I. Prigogine in [1], p. 3, 203, 461, 505.
[31] H. Reichenbach: The direction of Time. University of California Press, Berkeley 1956.
H. Reichenbach: The Philosophy of Space and Time. Dover, N.Y. 1958.
[32] L. Tisza in [1], p. 107, 206, 510.
[33] S. Watanabe in The Voices of Time (J. T. Fraser, Ed.). George Braziller, N.Y. 1966, p. 453.
S. Watanabe; Progr. Theoret. Phys. (Kyoto) Suppl. (1965), Extra No., p. 135.
[34] J. A. Wheeler in Battele Rencontres in Mathematics and Physics. W. A. Benhamin, N.Y. 1968.
J. A. Wheeler in [4], p. 90, 233.
J. A. Wheeler; Am. Sci. *56* (1968), 1.
[35] P. T. Landsberg: Thermodynamics with Quantum Statistical Illustrations. Interscience 1961.
P. T. Landsberg; Proc. Roy. Soc. *A 262* (1961), 100.
[36] J. Loschmidt; Wiener Ber. *73* (1876), 139 and *75* (1877), 67.
[37] E. Zermelo; Ann. Physik *57* (1896), 588 and *59* (1896), 791.
[38] I. I. Gringorten and A. J. Kantor in Handbook of Geophysics and Space Environments (S. L. Valley, Ed.). McGraw-Hill, N.Y. 1965.
[39] B. Gal-Or: Cosmological Origin of Electromagnetic Dissipation. To be published.
[40] W. J. Cocke; Phys. Rev. *160 (5)* (1967), 1165.
[41] R. Balescu in [1], p. 473.
R. Balescu; Physica *36* (1967), 433.
R. Balescu; Phys. Lett. *27A* (1967), 249.
[42] I. Prigogine, F. Henin, and C. George; Proc. US Nat. Acad. Sci. *59* (1968), 7.
[43] A. N. Cox: Stellar Absorption Coefficients and Opacities. In: Stars and Stellar Systems vol. VIII (G. P. Kuiper, Ed.). The University of Chicago Press, Chicago 1965, 195—263.
[44] H. Reeves: Stellar Energy Sources. In: Stars and Stellar Systems vol. VIII (G. P. Cuiper, Ed.). The University of Chicago Press, Chicago 1965, 113—193.
[45] D. W. Sciama: The recent renaissance of observational cosmology. In: Relativity and Gravitation (C. G. Kuper and A. Pepes, Eds.). Gordon and Breach, N.Y. 1971, p. 283.
[46] J. M. Blatt; Progr. Theoret. Phys. *22* (1959), 745.

ENTROPY CONCEPT IN PHILOSOPHY

Irena Szumilewicz

Philosophy is a persistent attempt to answer two fundamental questions: "What is the world like?" and "What is man's situation in the world?".

The aim of science as well as that of philosophy is to find answers to these two questions, and specially to the first one. Because of its methods and tools, and the desintegration into specialized branches, science can give neither a picture of the world as a whole nor explain the situation of man in the world.

With the development of science and its differentiation difficulties are constantly mounting. The development of interdisciplinary border fields – after a short time when a sort of integration may be noticed – produces new specialists though not unifiers but dividers.

Only philosophy with its synthetic character can give a picture of the world as a whole which can answer the very deep need of mankind to understand the world.

The role of science in material and spiritual life is becoming increasingly important. Thus philosophy searching for an explanation of the world, should watch the situation in science very carefully.

Philosophy means a critical attitude to the foundations of science, its methods and tools, its successes and failures, from a metascientific point of view.

When times are revolutionary and the foundations of science undergo revolutionary transformations, the philosophical criticism of science, and the philosophical solutions of some scientific problems may turn out to be very valuable for the position of the development of science. Such an example was the position of philosophy in the second half of the XIXth century, just before the revolutionary transformation of physics. Mechanicism suffered one disaster after another.

The old mechanical stereotypes were not consistent with new facts and theories. Physics and mechanistic philosophy underwent revolutionary changes. The second law of thermodynamics and its methodological and philosophical repercussions have played a special role in changing the Newtonian scientific outlook upon Nature. To show the role of entropy law in philosophy we shall give a short survey of the situation in science and its influence upon philosophy in the late decades of the XIXth century.

I. Mechanicism. The Main Sources of its Stability

Modern mechanicism is a philosophical trend which developed gradually form the beginning of the XVIth century. The end of the XIXth and the beginning of the XXth century was the period of the breaking down of mechanicism. In its mature shape mechanicism appears to have ontological, gnoseological and methodological versions.

The ontological aspect is founded on the supposition that in the whole reality the same laws being in the micro-, macro- and supermacro-spheres. The laws are without exceptions, they are independent of time-*everlasting*, and independent of space-*universal*. The world's structure was regarded as changeless and determined.

In the gnoseological and methodological versions - which are consistent with the ontological one, it is assumed that universal laws are adequately known: they are, namely, the laws of classical mechanics. According to mechanism Newton's laws will last for ever as the foundation of physics and of any other branch of science. The petrifaction of the mechanistic stage of scientific development, leads to certain consequences.

Processes in any field of physics used to be explained by reducing them to mechanics by using mechanistic models or by reducing them to terms, laws and theories of mechanics. After a short period mechanics appeared to be the basic domain not only of physics but of any other science as well. Mechanistic trends penetrated even into the fields of ethics and aesthetics.

The methodological directions of mechanicism especially in the branch of physics, were very detailed. The law of gravitation served as a pattern fcr any law, by its universality, the mathematical form of differential equation and the structure according to the model of central forces. In the law of gravitation the forces are central, and inversely proportional to the distance square. In comparison to the distance the masses were relatively small so they were treated as material points. The central forces acted in distance (actio in distans). No material mediator was needed.

The laws or theories which were not built according to the model of central forces were regarded as not perfect, as temporary. This was the fate of Fourier's theory of heat conducting.

"At last we see the aim of physics means the reduction of phenomenon of nature to unchangeable forces of attraction and repulsion, the intensivity of which depends on the distance only. The solution on this problems is the condition of a complete understanding of nature..."

According to central forces pattern the laws of physics should be time-symmetrical and, what follows, every process should be reversible. Until the first half of the XIXth century physical laws were in agreement with this requirement.

The scientific model of the world, contrary to the intuitive one, was time-symmetrical. According to human intuition time lapses in one direction only, from the past to the future, from birth to death. This gap between science and human intuition had its critical reflection in philosophy.

But apart from these difficulties mechanics has almost unlimited credit. Faith in the truth of Newton's law was founded on several points.

1. The role of mathematics

From ancient times mathematics was given the highest place among sciences. The increasing use of mathematics in modern physics has had great influence on its development. Its laws appeared at once to have an universal character: they were free from systematical errors. The scientific prognoses based on such laws were not only qualitative but also quantitative, and what is most important, they were easily conformed with experience.

Newton's laws in the form of differential equations if taken for granted, allow some special statements to be implied. The role played by these laws seems to be like the one played by axioms in geometry. Though the criterion of truth in empirical sciences is not the same as in geometry, it made no difference to the contemporaries of Newton because of their very weak methodological consciousness. So the glory that until then had surrounded mathematics was transferred to mathematical physics as well.

2. Successes in explanation and prediction

Faith in classical mechanics was caused by its successes in the explanation and prediction. The astronomical predictions were the most important. For example, the impression made by the prediction and discovery of planet Neptune, was so tremendous because it was related to the heavenly bodies. It is well known that astrology had previously been held in deep respect because it presented the movements of heavenly bodies as the direct expression of God's will which appeared in front of mortals. This admiration and respect has been transferred from astrology to physics and astronomy.

3. The institutionalization of science

The institutionalization of science and the increasing role of scientific authorities had great influence up on the popularization of common trust in the law of classical mechanics as unshakable and absolutly true. Afterwards, in the final decades of the XIXth century the institutionalization of science played a great role in the defence of Newton's mechanics when new scientific facts were coming out against the old high altars.

4. Philosophy

In the stabilization of mechanics a good deal depends on philosophy. Kant and the rationalists claimed that Newton's laws were a priori synthetic judgements, and the empiricists held them to be inductive generalizations based on hard empirical facta. The splendour conferred in virtue of philosophy accompanied the laws of mechanics almost till the end of the XIXth century.

The myths surrounding mechanics were the effects of many mutually related causes, and philosophy was the last but not the least of then.

Faith in the absolute validity of Newton's laws stimulated an optimistic approach to science and its possibilities. It inspired the scientific community in the search to explain all its related problems.

But in time, the same belief in the truthfulness of mechanics which was the stimulus of its successes, began to slow down the development of new ideas and impend the renewal of science.

At the end of the XIXth century the removal of mechanicism from science and culture became the conditio sine qua non of the further development of science. In preparation of this crucial revolution new trends occurred in philosophy, in the philosophy of science, in methodology as well as a trend of revolutionary transformation of the foundations of scientific knowledge. This extraordinary challenge prepared a way for a revolution in physics.

In preparing for the revolution in physics the entropy law and its philosophical repercussions played a crucial role.

II. The Second Law of Thermodynamics

The philosophical reception of the entropy law and its influence upon the development of physics is closely related with the history of physics. So few facts on this topic should be mentioned.

The second law of thermodynamics was discovered in 1824 by Sadi Carnot – a young French engineer. There are some mistakes in the first version because at that time Sadi Carnot was still an adherent of the flogiston theory. The first law of thermodynamics was discovered just 20 years later by Joule, Mayer and Helmholtz, independent of each other. Thanks to them and to the beautiful and clear considerations of Carnot's law by Clapeyron, and later those of Thomson (Lord Kelvin), Clausius was able, in the sixties of the XIXth century to clarify Carnot's errors and make them consistent with the law of energy conservation [5].

Clausius went further in his study of heat energy. He raised the question of the specific feature of the heat processes. They are irreversible. As a measure of irreversibility Clausius introduced a new function – *entropy*. According to Clausius any spontaneous process in an isolated frame is connected with entropy increase. He enlarged this conclusion upon the universe as a whole, supposing that such an extrapolation is permissible. Clausius expressed his conclusion in the form of two sentences (very much discussed):

– The energy of the world is constant.

– The entropy of the world tends towards a maximum.

The effect of Clausius' prognosis was tremendous. At last a law was found which made it possible to distinguish in a scientific way the future from the past. The second law of thermodynamics permits us to explain, in a scientific way, the thus far only intuitive impression that the world tends towards the direction from yesterday to tomorrow.

R. Emden [7] puts it very clearly. He wrote that in the gigantic factory of natural processes, the entropy law plays the role of a director who designs the courses of the processes; the principle of energy conservation plays the role of the bookkeeper, which takes care only of the balance of energy.

Ludwig Boltzmann intervented in defence of classical physics. He considered the protection of the old concepts to be the main aim of his life. Boltzmann supplemented mechanics with some additional premises. He added the presumption of equal probability of any microstate, and used the statistical approach. In his interpretation in an isolated frame which is not to be found in thermodynamical equilibrium, an entropy increase is not absolutely certain, it is only the most probable. But an entropy decrease is possible as well, there is only a very small probability that it will take place. This statistical entropy law is also known as Boltzmann's "H-theorem" [4].

It is exactly as follows:

If a closed system is in a non-equilibrium state of relatively low entropy, then the increase of entropy with time is overwhelmingly probable by virtue of the approach of the particles of their equlibrium distribution.

Contrary to Clausius' prognosis according to which the world tends to thermodynamical death, Boltzmann put forward his fluctuation hypothesis. The universe as a whole is in a state of thermodynamical equilibrium, and its entropy is constant and has reached a maximum.

Some subsystems with initial low entropy and some subsystems with initial high entropy branch out from universe as a whole. The increase (or decrease) of entropy depends on the initial condition of the subsystem. If we agree to define the direction of time as the direction of entropy increase, the universe as a whole does not exhibit a direction of time. However, some branch systems exhibit cosmologically "occasional" entropy decrease and temporally counter-directed ones, entropy increase.

According to Boltzmann some anti-entropic processes are possible. The entropy increase or decrease depends only on the initial conditions of isolated subsystems.

Soon after Boltzmann's research had appeared, Marian Smoluchowski the famous Polish scientist discovered, in the branch micro-events, some examples of anti-entropic processes (Brown's movements, critical opalescence, etc.).

Almost simultaneously Poincaré published his theorem. According to this theorem every process is reversible or quasi-reversible. Between the class of the co-called "reversible" processes and "irreversible" ones no short and fast lines exist. We call "reversible" such processes which have a short period of reverse, if compared with the observation time, and "irreversible" such processes which have a long period of reverse if compared with the time of observation.

On the basis of the above-mentioned scientific researches a very lively and fruitful discussion in science and philosophy has been started.

III. The Entropy Concept in Philosophy

The interest of philosophy in the entropy law was stimulated mainly by the following agents:

a) The entropy concept was closely connected to the philosophical interpretation at the very moment of its birth. Clausius not only formulated his law for isolated frames, but enlarged it for the universe as a whole, putting forward his almost apocalyptic prognosis about inexpiable heat deat of the world. This bold prognosis was far from the usual scientific caution.

b) The second law of thermodynamics put in a state of emergency not only the consistence of physics, but mechanistic philosophy as well. The model of central forces was shaken, so was the principle of actio in distans. Afterwards, when Boltzmann's statistical interpretation appeared the mechanistic pattern of a law and of the relation between a law and an accident were shaken as well. The mechanical outlook upon Nature has almost within a generation, been replaced, absorbed and transformed.

c) Shortly before the revolution in physics in the XIXth century, there was a sort of personal unity between science and philosophy. The top scientists were involved in philosophical and methodological discussions, and a few of them were founders of new philosophical trends. Mach was the initiator of empiriocriticism, Ostwald of energetism, and Poincaré of conventionalism.

d) Nowadays it is very well known that between science and philosophy there are no short and fast lines. This is especially evident in the periods of revolution and shortly before them. When the foundations of scientific knowledge undergo revolutionary transformation the philosophy of science receives an extraordinary challenge.

The entropy law put philosophy, to use psychological terminology, in a sort of cognitive dissonance. According to contemporary psychology, human consciousness when put in a situation of cognitive dissonance tends, through the processes of assimilation or accomodation to receive a consistent picture of the world.

Mutatis mutandis, one may observe in XIXth century philosophy tends towards assimilation of the entropy law through respective interpretations to one's philosophical concept. And one may notice tendencies towards the accomodation of the philosophical concept to the entropy law.

For example, the neotomistic philosophy assimilated Clausius' apocalyptic prognosis with great enthusiasm, in order to give new proof of God's existence. The Marxists used a quite different interpretation of the entropy law, in order to protect their materialistic point of view. The reciprocal effect of the materialistic interpretation upon science was productive, as will be shown when to the neotomistic view it was considered to be antiprogressive for the development of science.

Bergson and Ostwald accommodated their philosophical concepts to the consequences of the entropy law. Bergson constructed a new philosophy of creative evolution, antimechanistic and antiintellectual, in which the entropy law played a big role. Ostwald constructed, on the basis of the extrapolation of the entropy law, not only a new philosophy—*energetism*—but new physics as well. The approaches of Bergson and Ostwald were very fruitful and valuable for science, especially in their critical parts.

It is impossible – in a short article – to give an exhaustive description of all philosophical repercussions of the entropy concept. We intend rather to make a classification of the various attitudes with the main stress of their reciprocal effects upon scientific thought. The criterion of classifications is the relation to Clausius' prognosis. It seems to be the right one. In the XIXth century this problem was the very centre of scientific and philosophical discussion.

A. Adherents of the thermodynamical death of the universe

a) Catholic philosophy

Towards the end of the XIXth century one may observe that catholic philosophy came alive. It was the reaction to the stormy development of science accompanied by some sceptic and atheistic currents in philosophy. Catholic philosophy, like St. Thomas (see the encyclic of Leon XIII) searched to find some arguments for God's existence in science and especially in physics.

A new "entropy proof' of God's existence appeared on the basis of Clausius' prognosis. The first contour of the proof was drawn up by Adolf Fick in 1869, shortly after the publication of Clausius' research. The outline of this proof is as follows:

Fick tried to answer from where the Sun, whose energy is the only source of every process on earth, derives its energy. The answer based on Mayer's consideration is, that the main source of the Sun's energy is the gravitation energy of meteorites, their fall on to the surface of the Sun. "The spring of the world" is according to Fick the gravitional energy [9]:

"In der allgemeinen Gravitation hatten wir schliesslich gewissermassen die gespannte Feder gefunden, welche die ganze Weltuhrwerk in Bewegung erhält..."

Farther on Fick wondered if the development of the universe is cyclic or one-directed. Is it possible that the universe has come back to its initial conditions? The answer to this question is a negative one.

A spontaneous return to the starting point would be contrary to the entropy law. The universe as a whole represents a process tending towards an inexpiable heat death – towards a thermodynamical equilibrium. But, according to our observation, the universe is not to be found in a state of thermodynamical equilibrium which means that the world has not existed for ever but has a beginning in time. This beginning in time – the birth of the cosmos – means a state of minimum entropy. A state of minimum entropy could not be reached in a natural way, because all spontaneous processes are connected with entropy increase. The only explanation – according to Fick – is the creation of the universe by God. God not only created the universe in the sense that he minimalized entropy, but God simultaneously created all matter and energy. Neither the cosmos nor matter and energy has lasted for all time.

"The entropy proof" as we have seen, was based on an unlimited extrapolation of the entropy law to the universe as a whole. The proof starts with scientific considerations and ends by refuting the base of science – namely the laws of the conservation of energy and matter. An interesting polemic with neotomistic considerations has been started by Marxist philosophy [8] (see further on).

"The entropy proof" belongs to the class of proofs "ex inceptione mundi". As in other neotomistic proofs the material world is presented as dependent on God, and cannot exist without the act of creation.

The proof was accepted by neotomistic philosophy with great enthusiasm and afterwards repeated by many authors. Father Różycki wrote about the entropy proof:

"...it is a very succesful argument *ad hominem* for pupils devoted to science..." [19].

And really some scientists—for example the English astronomers Jeans and Eddington were adherents of the entropy proof.

Jeans wrote about the creation of the universe that the cosmos was like a clock the spring of which unwinds. According to Jeans the whole evidence with irrefutable force points to an event or to series of events creating the universe [11; 12].

Eddington also adopted the point of view of the creations of the universe. He wrote [6]:

"The beginning seems to present insuperable difficulties unless we agree to look on it as frankly supernatural."

With time, the statistical interpretation of the entropy law, the research of Smoluchowski, and Poincaré's theorem as well, somewhat weakened the categorical character of the entropy proof. In the new version of the entropy proof the heat death is described as not motionless but as a state without macroscopic processes. The trend towards thermodynamical death is an asymptote as stated by Ostwald. Heat death is no more certain, but only the most probable. But in spite of the changes the nerve of the entropy proof has remained the same since Fick's time.

The entropy proof was met by strong opposition and what is most interesting some catholic philosophers belong to its opponents. For example, the Polish priest Kazimierz Kłósak in his very interesting book "W poszukiwaniu pierwszej przyczyny" [13] evaluated the entropy proof as inconclusive from both scientific and logical points of view.

Evaluating the entropy proof from the historical point we may notice that, because of its anti-rationalistic edge and pessimistic approach to science, the proof stimulated some attempts to find arguments opposing Clausius' prognosis.

b) Bergsonism

Bergonism was one of the most impressive trends not only in French philosophy. The concept is an expression of disappointment with the situation in science with which Bergson was very well acquainted. One may say that Bergson raised his disappointment to the dignity of a philosophical concept. His criticism of mechanicism, which he identified with science, is very deep and keen. Science is the fruit of the intellect which is, according to Bergson, an inferior instrument of cognition while intuition is a superior one.

Intellect will never be able to give an adequate picture of the world. Only intuition can give the true picture.

Bergson analyses the mode of the working of intellect and in this way gives a devastating picture of contemporary science and its weaknesses. However his criticism goes beyond the criticism of mechanicism only.

Intellect is only a tool for pragmatic purposes, for the best biological orientation in the human environment. Intellect represents a system of perceptions assimilated for technical aims, it deforms the reality, it captures, from things and events which are individual, what they have in common what is repeated in them.

Through abstraction and analysis intellect isolates what is joined together, immobilizes what is mobile, simplifies and mechanizes what is alive and creative.

"Our mind is as long in its very element as it is under dead things." [2].

Time is a dimension for the intellect, like the space-dimensions. According to our mind every event, every process is reversible. There are no differences between the past, present and future.

Our mind is most helpless in the front of life and time. Only intuition which is a form of cognition, clean and disinterested, is able to give the truth. Intuition is an ability to identify itself with the real life of the world. The world is changeable and undergoes a creative evolution.

Some features of time, of creative evolution, are pictured by the second law of thermodynamics. According to the entropy law time is one-directed and irreversible. The fate of the world would be a thermodynamical death. Life which developed on the surface of the Earth is joined with inert matter, and though it may act contrary to the energy degradation, it cannot stop it, it can only delay the changes:

"...life... is an effort to pull up the load falling down... It succeeds only in delaying its falling down, but at least it may give an imago of what the pulling up of the load is like." [2]

Only God is able to stop the "falling down" of the world. God – in Bergson's interpretation – is a continuous act, it is nothing finished. God is freedom.

"...when I speak about a medium from which worlds squirt like skyrockets from a giant firework. I don't describe this medium as a thing but as an ejaculation." [2]

Man can identify himself with God through intuition which embraces the whole, the truth, not the isolated non-existing parts, as the intellect does. It is worth noticing that Bergson's God differs radically from the orthodox vision of God. His God is changing and creative and man is able to directly communicate with God through intuition without the intermediary of the Church. No wonder that Bergson's books were prohibited by the Church authorities.

Though it is impossible to accept Bergson's pessimistic irrationalistic and anti-scientific conclusion, his philosophy was useful for the development of science. Bergson's criticism of contemporary science and mechanicism of his time had no doubts about a reciprocal effect upon scientific thought and in this sense it was productive. Bergson's philosophy possibly unintentionally helped to replace and transform the naive mechanistic outlook.

c) Energetism

Energetism – a new philosophical trend which appeared in the second half of the XIXth century created by a famous scientist Wilhelm Ostwald – may be considered as a critical reaction to the difficulties in which the discovery of the entropy law places physics [16; 17].

Wilhelm Ostwald rightly understands that the moment has come for the foundations of physics to undergo revolutionary transformations. Mechanics was not consistent with the second law of thermodynamics. His diagnosis for this "serious illness" of science was to throw out mechanism and its fundamental term "mass" and replace it with the term "energy". The basis of new physics according to Ostwald should be the entropy law not the laws of mechanics.

When evaluating energetism historically it is evident that though Ostwald was almost right in the critical part of his concept, he was mistaken in the constructive one. The development of science depends not only on constructive propositions; without upsetting old altars, it is impossible to construct a new one.

Energetism proposed the construction of new physics and a new philosophy through the absoluteness of the entropy law as a fundamental law.

According to Ostwald [17] and his co-worker Holm [10] only various forms of energy exist. Any form of energy may be presented, like heat-energy, as a product of two agents: intensity and extensity. Intensity as in heat energy—*temperature*, and extensity as *the product of mass and specific heat*. Any spontaneous process takes place only if there is a difference between the agents of intensity. Every process in an isolated frame is joined with entropy increase, and lasts as long as the difference between intensity agents exists. The speed of the process is proportional to this difference. Every process tends to the annihilation of the intensity differences in an asymptotic way. At the beginning every spontaneous process is powerful and gradually it slows down.

"...erst gross und mächtig. Nun aber geht es leise, geht bedächtig." [10]

This is the fate of the world as well, claimed Ostwald. Thermodynamical death of the universe will be reached in an *asymptotic* way.

Ostwald was fascinated by the new picture of the world. The irreversibility of processes contrary to the reversibility according to mechanics is very important for mankind, wrote Ostwald. Not only physics changed but philosophy as well. Ostwald proclaimed that by removing the term "mass" he had destroyed the foundations of the materialistic trends in philosophy for ever. Between "energy" and "spirit" there are no differences. On this topic he read his famous paper "Überwindung der wissenschaftlichen Materialismus" in Lübeck in 1895. From then on he proclaimed [16]:

"...we may eliminate at last the horrible gap between matter and spirit which remained from the times of Descartes."

The irreversibility of the changes in reality are, according to Ostwald, specially important from the standpoint of morality and of any spiritual value as well. The responsibility for every act has sense only if the act cannot be repeated, if what is done is done for ever. According to mechanics every event, Ostwald wrote [17], could be repeated at any moment:

"Den jedes unwillkommene Erreignis liesse sich rückgänging machen und hiernach ein anderes Wünschenwertes an seine Stelle treten lassen."

With regard to the above cited sentence, one may wonder how a scientist could identify time symmetry of laws with the possibility of repeating any event in a free chosen moment.

Ostwald's philosophical as well as his physical concepts had some adherents but simultaneously met with very sharp opposition. The polemic between Ludwig Boltzmann and Ostwald was of special interest in his paper "Ein Wort der Mathematik an die Energetik" [3]. Boltzmann showed point after point many mistakes made by Ostwald. What is worse according to Boltzmann, Ostwald's proposition is not only full of scientific errors but it is not even simpler than classical mechanics. History shows that Ostwald was not right and if he had won his struggle against the atomists it would have damaged the development of science.

However, in the constructive part Ostwald was the loser, but in the critical one he was the winner. The reciprocal effect of his philosophy upon science was to show that mechanistic methods and stereotypes must be changed if the development of science was to continue.

B. The adversaries of Clausius' prognosis

a) Dialectical materialism – Marxism

Dialectical materialism is against Clausius' prognosis and against the implication which "the entropy proof" of God's existence has established. According to the methodological principles of materialistic philosophy Friedrich Engels rejected the explanation based on God's help. Engels accepted only the naturalistic explanations, so he rejected "the entropy proof". Engels [8] pointed out the scientific hiatus and logical inconsistency of the proof.

From the methodological point of view Engels raised the question as to whether the extrapolation of the second law of thermodynamics to the universe as a whole was legitimate. The enlarging of the entropy law is based on the supposition that every process in the cosmos is connected with entropy increase, that the universe as a whole is a process tending towards maximum entropy. But it is possible – wrote Engels – that the Universe undergoes a cyclic process, which has segments of entropy increase and segments of entropy decrease.

Engels hoped that if not the science of his day then the science of tomorrow would find the anti-entropic processes:

"...in a way which to find is a task of future science the heat emanated in the cosmic space, must have a possibility to transform into another form of motion in which it will be able to concentrate and to be active once more" [8].

The difficulties in the solution of this problem are connected – supposed Engels – with the fact that probably the periods of the cycle are very long if compared with the time of life of mankind.

From the methodological point of view, Marxism is against the petrifaction of the difficulties of present-day science. Science never deals with absolute truths, on the contrary, science knows only relative truths which are constantly improved according to the stage of development.

Engels pointed out the logical inconsistency of the entropy proof as well. The proof is based on the supposition that the laws of science are true and valid for the universe as a whole. However, the conclusion opposed the laws of conservation of energy and matter. If matter and energy were created they are facts shaking the universal validity of these laws. Engels rightly pointed out there was a logical error in the entropy proof.

The Marxist criticism of the "entropy proof" based on Clausius' prognosis was very fruitful from the scientific point of view. Engels questioned the extrapolation of the entropy law, its legitimate validity. That means a big blow to mechanicism. According to mechanicism, any extrapolations were legitimate. If the law of classical mechanics is valid in the micro-, macro- and supermacro-spheres, any law may be extrapolated in an unlimited way.

To raise the problem of extrapolation being legitimate means to shake the mechanistic naive supposition of the unlimited validity of laws in space and time, and in this way, to speed the transformation of the foundations of physics. The materialistic view was opposed to the supernatural explanations which means that this philosophy defended the very basis of modern science and its independence of theology.

b) The theory of the endless return of worlds

The idea of cyclic changes of the universe has been raised not only by Engels but by a group of philosophers and scientists, the adherents of the theory of the endless returns of worlds.

The idea is almost as old as philosophy itself. In ancient times the stoics were its adherents and in modern times its followers were August Blanqui, Friedrich Nietzsche, Svante Arrhenius [1], Zygmunt Zawirski [20], Abel Rey [18], etc. The idea of the endless returns may be interpreted as an attempt to solve the tragic dilemma of man's life, to mitigate the fear of death. The modern adherents of this theory base their scientific arguments mainly upon Poincaré's theorem. The theory of the endless returns presupposes long cycles so it is very difficult to find some empirical evidence for the theory. The Swedish scientist Svante Arrhenius, however, made an attempt to find some segments of the cycle where entropy increases and others where entropy decreases. The increase of entropy proceeds through radiation of heat to cold nebula, the decrease through the departure of the fastest molecules and their condensation in the compact concentration of matter [1].

The theory with its optimistic approach, however interesting, was too speculative to have any visible influence on the development of science.

c) Empirio-criticism

Empirio-criticism associated with the name of Ernst Mach was a radical form of empiricism, its policy was to base science upon hard facts only. The failures of science are caused, according to Mach, by the anthropomorphic and metaphysical terms which adulterate science. To eliminate from science these kinds of terms was Mach's main object. For example, "absolute space", "absolute time", "cosmic ether" and the causal principle [14; 15] – the fundamental terms ofmechanicism – are not empirical terms.

It is impossible to measure absolute time at all. The physicist measures the relative time only comparing the experienced time periods with clocks.

"The absolute time has neither practical nor scientific value. Nobody has the right to say that he knows something about it. This is a barren metaphysical term."

Mach applied the same kind of criticism to absolute space and cosmic ether. All the terms are metaphysical. The causal principle was condemned as antropomorphic. Mach's criticism was markedly against mechanicism. He wrote it expressis verbis [15]:

"When the mode of description is so limited and non – elastic that does make it possible to follow the multisidedness of phenomena it is impossible to use such a formula, it begins to hinder the cognition of events... I suppose that this takes place with the mechanistic approach to physics."

According to the policy of radical empiricism Mach expresses his opposition to the extrapolation of the entropy principle on the universe as a whole. One may imply from the context of Mach's considerations that he accepted only as legitimate such extrapolation that is not beyond the scope of possible experience. But it is impossible – wrote Mach – to experiment with the whole cosmos. It is impossible to measure its entropy or energy, so Clausius' prognosis that the world tends towards thermodynamical death is *unscientific* [15]:

"Es sei noch gestattet zu bemerken, dass die Ausdrücke 'Energie der Welt' und 'Entropie der Welt' etwas von Scholastik an sich haben. Energie and Entropie sind Massbegriffe. Welchen Sinn kann es haben, diese Begriffe auf einem Fall anzuwenden, auf welchen dieselben eben nicht anwendbar, in welchen baren Werte unbestimmbar sind?"

The further development of the philosophy of science and especially of methodology made clear that Mach's fundamental idea is not only disastrous for the development of science, but simply and for many reasons impossible to realize. Science cannot be founded upon hard empirical facts – simply because such facts do not exist. But Mach was absolutly correct in his criticism of mechanicism. Mechanicism "hinders the development of science". Mach's struggle against mechanicism was, no doubt, a big contribution to the development of science. The same may be applied to his considerations on extrapolation. To raise the problem of extrapolation legitimacy means to break the old mechanistic sterotypes, according to which, the validity of laws is unlimited.

IV. Conclusions

We may sum up our conclusions as follows:

It has been impossible to describe adequately in a short paper the philosophical repercussions of the entropy law. Our aim is more modest. It is only the clasification of various attitudes with the stream of their reciprocal effect upon the development of science.

One may notice two generally different classes of philosophical attitudes to the difficulties in which science was placed by the discovery of the entropy concept.

Firstly. Devotion to science, a desire to help out science with the criticism of mechanicism, of the old foundations of sciences and to offer science some new ideas.

And secondly. A counter scientific attitude, a distaste for science (e.g. Bergsonism). It may seem that only the first class of attitudes was useful for the development of science. Such an evaluation would be superficial. Science needs criticism and praises it very highly. For example Bergson with his deep and keen criticism of mechanicism helped in the development of science. In spite of his (alleged) anti-intellectualism he showed in a very impressive way, that science, progressing by isolating bits of broader experience, never seeks to grasp the whole except mystically – intuitively. In this way he raised a new aim for science to find how to construct the whole from isolated parts.

The other attitude devoted to science offers several benefits. The radical criticism of mechanicism with its absolute time and space with its cosmic ether and model of central forces (Mach, Engels, Ostwald) helped a great deal to understand that the old stereotypes must be replaced and transformed. To awake the problem of extrapolation in physics its limits and borders was very useful for the development of the methodological consciousness of the epoch. In effect, it stimulated a discussion about the relation between empirical evidence and laws and theories.

The philosophical reception of the entropy concept put forward, in connection with the question of extrapolation, some cosmological problems such as the space and time structure of the universe, and the character of the development of cosmos (cyclic or one-directed).

Even the theory of the endless returns so fantastic and speculative, almost like science fiction, was in a way a benefit to science. The theory encouraged the scientist to look bravely for new solutions to new ideas.

The development of science depends not only on accuracy, clarity and criticism but on courage and fantasy as well.

References

[1] S. Arrhenius: Obraz wszechswiata w dziejach ludzkości (Worlds in the Making. The Evolution of the Universe). Warszawa-Lwów 1912, p. 147.

[2] H. Bergson: Ewolucja twórcza (L'Évolution Créatrice). Gebethner-Wolff, Warszawa 1913, pp. 210, 212.

[3] L. Boltzmann: Populäre Schriften. Johann Ambrosius Barth, Leipzig 1905, pp. 104—140.

[4] L. Boltzmann: Vorlesungen über Gastheorie. Leipzig 1896—1898.

[5] R. Clausius: Über die bewegende Kraft der Warme und die Gesetze welche sich davon für die Wärmelehre selbst ableiten lassen. Poggendorfs Annalen, 1850.

[6] A. Eddington: The Expanding Universe. University Press, Cambridge 1933, pp. 125.

[7] R. Emden: Why do we have Winter Heating. Nature *141* (1936), 908.

[8] F. Engels: Dialektyka przyrody (The Dialectics of Natur). KiW, Warszawa 1953, p. 26.

[9] A. Fick: Die Naturkräfte in ihrer Wechselbeziehung. Populäre Vorträge B. I. Würzburg 1869.

[10] Helm (Geo): Die Energetik nach ihrer geschichtlicher Entwicklung. Veit & Co., Leipzig 1898.

[11] J. Jeans: Horyzonty nowej nauki (The New Background of Science). Mathesis Polska, Warszawa 1937.

[12] J. Jeans: Nowy świat fizyki (The Mysterious Universe). Trzyska, Evert i Michalski, Warszawa 1932.

[13] K. Kłósak: W poszukiwaniu pierwszej przyczyny. Pax, Warszawa 1955.

[14] E. Mach: Die Prinzipien der Wärmelehre. J. A. Barth, Leipzig 1919.

[15] E. Mach: Populäre wissenschftliche Vorlesungen. J. A. Barth, Leipzig 1910.

[16] W. Ostwald: Überwindung der wissenschaftlichen Materialismus. Veit & Co., Lübeck 1895.

[17] W. Ostwald: Die Philosophie der Werte. A. Kröner, Leipzig 1913.

[18] A. Rey: Le retour eternel et la philosophie de la physique. Paris 1927.

[19] I. Różycki: Dogmatyka II — Istnienie Boga. Skrypt autoryzowany nr. 230. Kraków 1948.

[20] Z. Zawirski: Wieczne powroty światów. Kwartalnik Filozoficzny, Kraków 1927.

INFORMATION, KNOWLEDGE AND TIME

Jiří Zeman

I.

The modern concept of information has considerably influenced the philosophical problems and has some important consequences for epistemology and for ontology, too. Information theory and cybernetics threw fresh light on such concepts as knowledge, time, evolution etc.

If we want to illuminate the essence of the information principle in human cognition we must investigate the epistemological characteristics in general, we must determine the so-called epistemological situation, the situation of the human observer in the universe and their parameters. An analysis of this situation provides the basic conditions of cognition which are actually the limiting conditions. The epistemological situation cannot be described in a wholly exhaustive manner – it is necessary to reduce it to its basic parameters, to construct its model. Each formulation is made in some language which uses certain concepts, principles and relations; some of them are basic, starting. A certain description differs according to the initial presumptions; it even differs in the individual sciences. The task of philosophy is to try to give a general, universal description of the epistemological situation, to determine its parameters in general and thereby to establish actually the relative limitation of cognition, the relative degree of our lack of knowledge – for the purpose of overcoming it in the further development. Information theory brings new stimuli to this investigation. The human observer[1]) lives in the world which has certain space-time and other characteristics. He finds himself in a certain area of epistemological accessibility from whose parameters he must begin his investigations. He lives in the macro-world – between the megaworld and the microworld, he lives in a sphere where certain laws are valid, he is relatively limited by his senses, intellectual abilities, experience – he is limited by the properties of his environment and by his internal properties,

[1]) By the word "observer" I mean the human, epistemological subject, the information receiver, e.g. the scientist. I am considering here a part of the epistemological process only and I am not trying to give an exact analysis of the practical activity which—as the basic component of it—is very complicated (in relation to social and historical conditions) and the modelling of it would be too difficult and simplified.

too. His empirical observation and logical thinking do not have unlimited precision. The totality of human knowledge is based on the finite number of acts of observation and thinking, the amount of information contained in them is finite. Human knowledge has a macroscopic and statistical character since it proceeds from the biological and psychical attributes of man and has a particular degree of uncertainty (imprecision). Knowledge is dependent on the human information capacity: human information can never exceed the capacity of the epistemological channel although it can increase with the development of cognition.

The task of epistemology is a detailed examination of this capacity.[2]) The supposition for an act of cognition to arise is the existence of a brain with the ability of consciousness and memory; furthermore, a source of physical energy (for instance light), a source of physiological energy conditioning the emergence of excitational processes and an objectively existing source of information. There exist certain logical principles of epistemological transformation which condition the character of cognition. Modern science – in contrast to the former "classical" science – is approaching the point of discovering certain constants or principles that determine particular marginal conditions valid in scientific cognition. There are, for instance, the constant of the light velocity, Planck's constant, Boltzmann's constant, Heisenberg's relation of uncertainty etc. Finally, it should be noted that all processes – including the cognitive one – are dependent on certain characteristics of the universe such as its anisotropy, orientation etc. Subjective, epistemological characteristics are thus subordinate to objective, cosmic parameters. For instance, experiencing time and recognizing time occur on the basis of an objective time flow existing in the universe; they are dependent on consciousness, whose psychical processes arise on a biochemical and physical basis. Thus the subjective and the objective intertwine and the task of epistemology is to examine this relationship in detail. The subjectivity of man is linked to his spatial size, located between the ultralarge and the ultrasmall, and on his time dimensions, characterized in the same way. The picture of the world also depends on the length of our "momentum", i.e. on the time character of sensual and psychical experiences and reactions. We are not always fully aware of these facts. We often forget that information proceeds from the information flow, which is dependent on the cosmic flow.

Albert Einstein expressed the thought that an intelligent mechanism could not – probably – work in a balanced environment. Such an environment in equilibrium might be, for instance, the universe with either too great or too thin density of matter and energy, which is to say the universe where the oriented processes would not occur. The universe is an environment in which gravitational, electromagnetic and thermodynamic processes are at work. In order that these processes happen there must exist a certain asymmetry in the universe – an imbalance, which permits the flow of processes having an orientated character and an irreversible gradient with a given production of entropy. Against the background of this cosmic gradient

[2]) The field of epistemology is very broad and includes not only the aspects of natural sciences but also the aspects of social practice, of the social-historical conditions of the epistemological process, of the class and economic roots of sciences etc. These aspects are not considered here. I have taken into account the very narrow part of the epistemological process—the possibility of the description of conditions in the transfer of information from the objective source to the human receiver on the basis of certain exact approaches.

(or to put in other words, in the framework of the cosmic channel)[3]) other, non-physical processes take place—chemical, biological and psychical ones. The principle of life, the principle of thought, information processes cannot go on without gradient, without imbalance, without differences in certain levels. The living organism can only exist on the basis of the flow of matter and energy which are degraded in order to maintain their level of organization. The information processes too (conscious, intellectual) assume a flow (of matter, energy and information) and, ipso facto, a degradation. Observation is an irreversible process based on the elementary differences between two states of time, two states of energy and two states of information. Every appearance of thought and its expression also includes, apart from information gain, energy and information degradation; the moment of depreciation (that is of loss) must be examined among all products of thought and cognition. No information system can function without producing entropy (physical and informational). At the same time, the information acquired represents a fund which is always at science's disposal without having to wrest it from nature each time and with great effort. Aside from the element of depreciation the results of cognition also include the element of effectiveness (work, or better said memory effect). Informational entropy expresses the rate of ignorance about a phenomenon, i.e. the rate of information depreciation or limitation. The entropy of knowledge (in the form of a sensual-empirical or intellectual-logical) is never zero. The information process, from the physical aspects, is a working process—just as is every process of non-living nature—and it assumes a gradient caused by some imbalance. The working capacity of the information system is always limited and is established by the coefficient of the effectiveness of a system. Information is the result of a working process of the information system. The brain, endowed with memory, is able to raise the coefficient of its efficiency. As for informational and physical entropy, from the philosophical viewpoint it is possible to assume their similarity (if not total identity), and this view is upheld by a number of scientists.

The existence of an oriented and irreversible process in the universe can be determined empirically. The cosmic flow is manifested by the expansion of the universe, the flow of electromagnetic rays, by entropical processes. The universe can be understood as a dynamic system or channel in which one can, at a given moment, determine the average total speed of flow, the flow capacity and the relative density of matter. All other processes occur on the basis of this cosmic flow and they are also oriented and irreversible, they have a certain speed and capacity of flow and are linked with a certain density of a given system (for instance, matter, energy, information). Both the cosmic and the information flow assume an imbalance, irreversibility, anisotropy and heterogeneity. In the universe, there are proceeding processes of the dispersion of matter, energy, heat, processes of increasing entropy and overall dilution. Against the background of this disorganizational, degradation process there are, however, coalescent processes—processes of condensation, concentration; there is a growth of organiza-

[3]) I do not support here any conception of physicalism. The general physical laws are valid on all of the higher levels of motion; e.g. the properties of gradient, orientation, direction etc. are applied here, too. The chemical, biological, psycho-social motions include the physical motion, too, but at the same time they also represent some new higher qualitative degrees over this physical motion; I am not concerned here with the proper characteristics of these motions.

tion or information in the form of life and learning. The conservation ("mnemic") element leads to the retention of the working effect (in the form of organization or information). Cognition assumes the issuance of negentropy, but also leads to its partial maintenance. Life, learning and knowledge happen on the basis of cosmic flow and are subordinate to the principle of entropy; they are, however, also linked to the flow in the opposite direction, i.e. processes of condensation and the increase of negative entropy.

The rate of information density can be expressed with the aid of redundancy as

$$R = 1 - \frac{I_{\text{ef}}}{I_{\text{max}}},$$

where I_{ef} is information effective, acquired, and I_{max} is information maximum, emitted. This rate is given also in a certain relation to the capacity of the information system. The parameter of information density has not yet been taken into account in epistemology and logic, although this is a very important thing relating to the question of epistemological transformation.

The world as a whole is a system that is more complicated and richer than its image in cognition, i.e. it is different from it. The information capacity of a human being is relatively limited, therefore the world is only reproduced in a simplified form, and there is a loss in the process of cognition. The limitations of the epistemological capacity proceed from the relatively limited neurophysiological and psychical possibilities. The performance of the brain depends on the characteristics of its structure and dynamics; consciousness represents straits which carry out a certain form of selection; the output capacity also relates to the individual senses, memory, thinking. The overall scientific image of the world depends on these factors linked to a level of science and its methods, together with the development of technology, of scientific instruments, etc. The obstacles that stand in the way of the cognitive process in the transition from object to subject can be summarily called, according to information theory, noise (epistemological noise). The performance of noise has an influence on the result of the information process and leads to the constriction of the initial variety of originals (variety of emitted information). The cognitive process does not take place passively, so that noise makes the choice instead of the subject. The subject (receiver) has the ability to actively choose what it needs from a vast amount. Thus the limiting of variety has its negative and positive aspects, its passive and active moments; objectively existing noise does not facilitate access to all information about the world, it causes uncertainty in the image and has an influence on factual information output; simultaneously, the subject itself performs the selection, actively limiting the variety and choosing what is important, what is basic from what is accessible (the area of inaccessibility is, in fact, determined by the noise!). Thus the complexity and variety of relations is reduced in cognition (simplified, impoverished, coarse-grained). The individual concepts and the total image of the world are drawn in outline, certain aspects of the overall context are isolated and removed from the flow of time. The continuous is transformed into the discontinuous, the indivisible into the divisible, the total into the partial. Epistemological transformation means simply transformation into a system of other relations and principles. This is a very complicated process and its detailed analysis would be a very broad and challenging research job. On the whole it can be said that in the macroimage of the world, as our senses, reason and science present it, parts represent the whole, partial information about the world represents the total; it is the consequence of the noise and of the limited epistemological capacity. Thus information loss takes place here; uncertainty (entropy) in the image means information dilution or this image is redundant—metaphorically speaking it is as though it were "stuffed with cotton wool". Information degradation is always established even though it is reduced in the development of cognition. It proceeds from the division in the objective and the subjective, in the original and the image. Problems of noise, redundancy and information capacity relate to the time question. The information flow has time characteristics (performance of the channel is related to time) and these even have the memory which the acquired information reflects and retains (retention of the effect of the working process). Effective information is always less than the maximum possible information, or to be more precise, it is smaller, or in ideal instances, it is at best equal to maximum information minus a certain elementary increment of information loss:

$$I_{ef} \leqq I_{max} - \Delta I_{lost}.$$

The elementary amount of loss influences epistemological transformation and, thereby, the character of the image of the world. The performance of cognition is related to the performance of consciousness; its effectiveness is never the optimum, there is always a certain degradation, i.e. a certain information loss. The information performance and the noise effect influence the resulting forms and content of our sensual perception, logical thinking and total picture of the world. The information effect is established in relation to subjective conditions (the character of knowledge, the brain, etc.) and objective conditions (the laws and character of the cosmos, the cosmic flow), although both are not separate from each other. These are the conditions that limited the information zone, the sphere of epistemological accessibility which is not entirely invariable but in the given confines is always unsurpassable.

II.

The zone of epistemological accessibility is established, on the one hand, by the attributes of the objective world, and, on the other hand, by the subjective characteristics of the human observer. The sphere of cognition is restricted to the dimensions and attributes of human beings. The macroworld is the starting point through which man penetrates the megaworld and microworld; all these fields are mutually related in a certain logical way. Human cognition is confined to the space and time characteristics of man; as regards time there is, on the one hand, a certain tempo of biological and psychophysical processes, and, on the other hand, the length of life of an individual and of social history. Man starts from characteristics and laws of sensual cognition and logical thinking, which are improved with the aid of instruments, scientific methods, etc.

The information sphere of man is further restricted to a sphere that has a certain character and direction of physical (for instance thermodynamic) processes. For instance, we would not be able to communicate with a system that has a contrary direction of electromagnetic and thermodynamic processes (cf. hypothesis of the so-called black stars swallowing light). As was already mentioned, information processes cannot take place in a state of equilibrium. They must have a certain direction, which depends on the cosmic flow. The cosmic flow establishes the character of the physical, and proceeding from this, even chemical and biological, processes. The cosmic and information flows have a certain direction, speed and capacity. Cognition can take place only in the frame of this flow. In the framework of the cosmic flow are given the objective time, space, physical and chemical attributes of the world, from which follow the biological and neurophysiological characteristics, and from these then come the information characteristics and laws of subjective cognition. The human observer is placed in the flow of cosmological channel and, simultaneously, he has a special ability of reflection and of memory. Thus, the information sphere of man has certain parameters, a relative limitation which should be precisely examined since they are the point of departure for determining the basis of science, the laws of cognition. The character of depiction, the character of the information systems, after all, depends on determining the initial parameters.

Thus, for instance, in classical physics the universal constants were actually simplified to infinity (e.g. the speed of light c as ∞) or zero (e.g. Planck's constant h as 0). If the speed of light approaches infinity, the isotropic area of the quasi-simultaneity in the Minkowski's time-space is completely flattened, falls out, and with it the whole system of relativistic phenomena. This is exactly the case of classical physics [1, pp. 87–88] (Fig. 1).

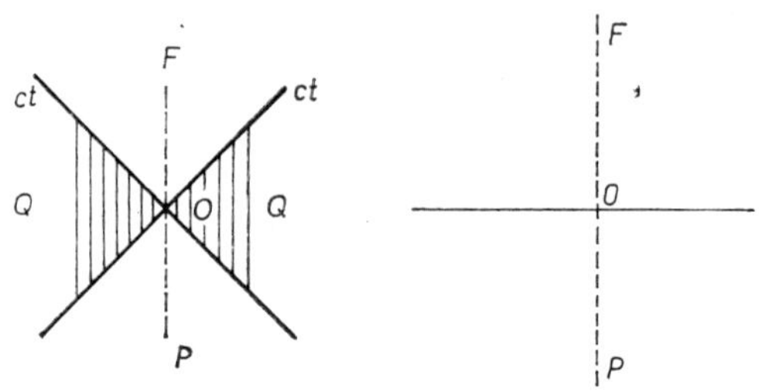

Fig. 1 (F — future; 0 — present; P — past; Q — quasi-simultaneity; c — speed of light; t — time).

For instance, the manner of depiction on maps depends on the initial manner of the design. Euclidean geometry did not take into consideration the parameter of crookedness of space and was therefore applicable only in the special case of flat geometry as distinguished from non-Euclidean systems.

According to Brillouin the relation between the increment of entropy and of information can be expressed as

$$\Delta S = k \ln 2 \, \Delta I,$$

where k is Boltzmann's constant. In the so-called precybernetic stage of physics, according to Costa de Beauregard, this k approached zero, or observation—the transfer of part of the negentropy of the system into the information of the observer—was carried out without cost—that is without an increment of entropy. In this case the parameter of information loss was not taken into consideration, the losing amount was like zero and real cognition was, so to speak absolutely precise.[4])

The limitation of information capacity means that the transmitted information is distributed and extended in time, i.e. it does not go from the past to the future by the infinite speed. If we accept the speed limit is the speed of light c we can *connect the relativistic model of the world with the model of information channel:* the transition of information from the source to the receiver is the transition from the past to the future going by a certain speed.

Very slow information transmission could be conceived as connected with a great redundancy (R approximating to maximum) and the maximum information speed connected with a very small redundancy (R approximating to minimum) (Fig. 2).

Fig. 2 (v_{min} — minimum speed of information transmission connected with $R = \max$; v_{max} — maximum speed of information transmission connected with $R = \min$; r_{inf} — sphere of information receiver; s_{inf} — sphere of information source; Q — sphere of quasi-simultaneity (normal principles of information transmission are not valid here, e. g. information loss)).

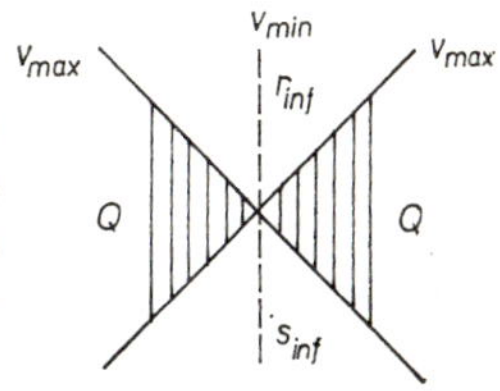

On this basis we can then get the following table of epistemological subjects (information receivers) with different capacity of transmission:

		S_1	S_2	S_3
v	(transmission speed)	min	medium	max
R	(redundancy)	max	medium	min
S	(entropy, performance of noise)	max	medium	min
I_{ef}	(effective information)	min	medium	max

Information uncertainty can be expressed also by the degree of information dilution in a message—by redundancy. In order to obtain certain truthful information certain work must be performed: the process of observation or learning requires activity whose results are then retained in the memory so that the activity need not be repeated. This is how the information content in cognition, i.e. in all scientific cognition, is condensed. The content of cognition is linked with the past. Cognition—message expressed in the form of concepts, sentences, laws—can contain various amount of information or has various degree of information density (i.e. dilution—redundancy). The pre-cybernetic stage did not take into consideration either the positive increment of entropy in every act of cognition or thc positive increment of in-

[4]) Heisenberg for instance noted the limitation in the preciseness of the physical determination of the microworld; Gödel pointed out the limitation of mathematical formalization.

formation density in the result of the act of cognition. Various concepts (regarded as epistemologically relevant), as well as various information systems (scientific systems), placed alongside one another do not have the same degree of redundancy. This is evident, for instance, from the fact that two sentences or two articles speaking about the same subject and using the same amount of symbols, need not have the same information density. The growth of information density, established by memory, i.e. the development of cognition, can be depicted for instance in Minkowski's time-space, where we would add the minimum speed of the information process to maximum redundancy and the maximum speed of the information process to minimum redundancy.

Logic has not yet taken into account the information content of concepts and statements, or the parameter of information density and also, of course, time; this logic deliberately abstracts from the content. But the very question of so-called semantic information, dealing with the meaning of messages, cannot leave aside this viewpoint. All the problems indicated above would require attention and extensive examination; in this contribution we can merely call attention to them briefly.

III.

Information is in certain relation to time, too. For the processes of progress (increase of information, negentropy, order) or regress (increase of entropy, disorder) the non-linear metric is more suitable than the linear metric. Intervals of the mechanical (physical, astronomic) time are homogeneous – the time flow is here given as uniform. The measure of this time are, for instance, certain mechanical processes (e.g. rotations of Earth) repeated in the same time intervals: the intervals of the past and of the future are homogeneous, are not shortened or lengthened. The scale of the mechanical time is linear. The entropic or negentropic (informational) processes (processes of decay, development, learning) can be conceived as processes of successive dilution or condensation. The physiological time of Lecomte du Noüy (healing of wounds), the biological time of Gaston Backman (biological growth) etc. can be measured by the changing amount of physiological or biological events. These times are non-linear. Also the so-called atomic time (or Milne's kinematic time) is non-linear: for measuring the processes of radio-active decomposition the atomic clock is more suitable (the half-time of the radio-active decomposition is constant in the atomic, kinematic time only but is not constant in the mechanical time).[5]

[5]) According to G. Frey the relation between mechanical time t and atomic time T is given as

$$t = \frac{1}{\lambda} \ln T.$$

λ is here the constant of the radio-active decomposition.

Non-linear time can be represented by the exponential or logarithmic curve. In relation to the constant intervals (with constant rate of events-units) in linear, mechanical time the rate of events (units) in non-linear time is changing (processes of dilution or condensation). The time can be measured also by information (negentropy), e.g. in the case of learning, development of knowledge, growth of order etc.

Informational (negentropic) time is also non-linear and can be measured by increasing, accelerating, condensing information content (or amount of negentropy). The process of learning (in brain, society, nature) means in general a development with the accumulation of information. This process can be 1. slowed down [a) to a maximum, b) in infinity] or 2. accelerated [a) to a maximum, b) to infinity].

We mentioned also the possibility of the connection of relativistic world and information transmission. We can imagine the speed of transmission (with the limit of the light velocity c) as correlated to information density (some elementary information amount corresponding to some elementary transmission speed, maximum information amount corresponding to maximum speed). The acquisition of information needs certain time and certain expenditure of energy. The acquired information is the potential, stored information given in the memory without a necessity of this expenditure. The process of the re-using of this information (e.g. recollection) can be conceived as a certain kind of reading, as a very fast information transmission. Redundancy is connected with a certain slowing-down of transmission; accumulation of information is connected with a condensation of information, diminishing of redundancy, acceleration of transmission.

Relativistic phenomena are connected with very fast velocities. We can imagine that e.g. certain high organized brain phenomena (re-using complicated information) are related in fact to very fast transmission ("fast reading": the student of 20th century can learn in a certain

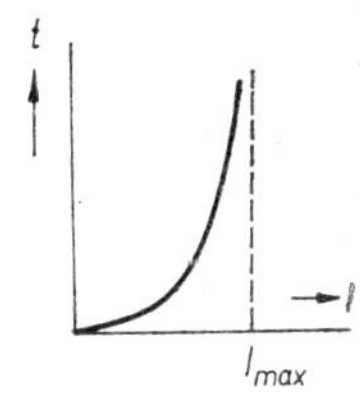

Fig. 3 (t — time; I — information amount (connected with certain transmission speed); I_{max} — maximum information amount (connected with maximum transmission speed)).

time more condensed information than the student in the middle-ages). If I_{max} (maximum condensed information connected with maximum transmission speed) corresponds to speed c, we can get the following diagram of the relation of time t and information I (Fig. 3). This relation can be expressed by the formula

$$t = -\tau \ln \left(1 - \frac{I}{I_{max}}\right),$$

or

$$t = \tau \ln \frac{1}{R}.$$

τ means certain time constant, R means redundancy. Time is expressed as the negative logarithm of redundancy (or as the logarithm of inversed redundancy). The case of maximum information is related to maximum dilatation of time.

The memory can be represented as a certain kind of time-space (informational time-space). Further, the memory as an accumulated quantity of information is in certain relation to redundancy.

According to the materialist theory of reflection the matter has the general property and ability of reflection, of information record. We can differ between so-called structural (internal, bound) information (or: structure of a system) and so-called signal (external, free) information (or: message). The relation of these two kinds of information can be expressed by the rate of density (high-low). Signal information can be condensed into the structural information. The higher organized system can be characterized by the higher degree of stored, accumulated structural information. The structural can be related to the condensed, stored, composed in time; the diluted, dispersed, extended in time is the opposite. The structure of an organized system represents in a certain sense the condensed record of the time flow (or: the conserved, effective part of a certain spent work):

The analysis of the concepts of information, time, redundancy, energy, record, reflection etc. (and of the relation between these concepts) is the basis for the monistic conception of the world, of the hierarchy of qualitative degrees of matter created in the evolution as the accumulation of negentropy and information.

References

[1] O. Costa de Beauregard: Le second principe de la science du temps. Editions du Seil, Paris 1963.

[2] J. T. Fraser, F. C. Haber, and G. H. Müller (Eds.): The Study of Time. Springer, Berlin-Heidelberg-New York 1972.

[3] G. Frey: Gesetz und Entwicklung in der Natur. Rowohlt, Hamburg 1958.

[4] Л. А. Петрушенко: Самодвижение материи в свете кибернетуку (L. A. Petrushenko: Self-motion of Matter in the Light of Cybernetics). Наука, Москва 1971.

[5] G. E. Shannon and W. Weaver: The Mathematical Theory of Communication. University of Illinois, Urbana 1949.

[6] A. D. Ursul: Information. Dietz, Berlin 1970.

[7] N. Wiener: Cybernetics. Hermann, Paris 1948.

[8] J. Zeman: Poznání a informace (Knowledge and Information). NČSAV, Praha 1962.

[9] J. Zeman (Ed.): Time in Science and Philosophy. Academia and Elsevier, Prague—Amsterdam 1971.

ABOUT THE AUTHORS

PAUL C. W. DAVIES, PH.D., is working as theoretical physicist at the Department of Mathematics, King's College, London, U. K. Born in London in 1946, he obtained B.Sc. and Ph.D. degrees at University College London, Physics Department, where he worked on problems in atomic physics and quantum electrodynamics. He developed an interest in cosmology, and in 1970 went to work as a Visiting Fellow at the Institute of Theoretical Astronomy in Cambridge, on topics in cosmology and relativity; in particular the study of the nature of time. In 1972 he returned to London as a lecturer at King's College, where he is at present engaged on work in cosmology and basic quantum theory.

Main publications:

The Physics of Time Asymmetry. University of Surrey Press (in prep.).
Charged Particle Creation in Cosmology. Il Nuovo Cimento *6B* (1971), 164—178.
Time Variation of the Coupling Constants. J. Phys. A *5* (1972), 1296—1304.
The Thermal Future of the Universe. M.N.R.A.S. *161* (1973), 1—5.

OLIVIER COSTA DE BEAUREGARD, PH.D., is the Director of the Centre National de la Recherche Scientifique, Laboratoire de Physique Théorique, Institut Henri Poincaré, Paris, France. Born in Paris in 1911, he mainly worked in theoretical physics (relativity and quantum theory) and has numerous publications in the philosophy of science.

Main publications:

La Théorie de la Relativité Restreinte. Masson, Paris 1949.
La Notion de Temps, Equivalence avec l'Espace. Hermann, Paris 1963.
Le Second Principe de la Science du Temps. Editions du Seuil, Paris 1963.
Précis de Mécanique Quantique Relativiste. Dunod, Paris 1967.
Précis of Special Relativity. Academic Press, New York 1966.

RICHARD SCHLEGEL, PH.D., is Professor of Physics at Michigan State University, East Lansing, Michigan, USA. Born in Davenport, Iowa (USA) in 1913, he became B.A. at the University of Chicago in 1935, M.A. at the University of Iowa in 1936 and Ph.D. at the University of Illinois in 1943. He studied in philosophy, physical chemistry and physics. He worked as associate physicist at the University of Chicago (1943—1945), as instructor of physics at Princeton University (1945—1948) and as professor of physics at Michigan State University (till now).

Main publications:

Time and the Physical World. 2nd ed., Dover Publications, Inc., New York 1968.
Completeness in Science. Appleton—Century—Crofts, New York 1967.

Inquiry into Science. Doubleday, New York 1972. (First published in a German translation: Steckbrief der Wissenschaft. Deutsche Verlags-Anstalt, Stuttgart 1969).

The Finite Universe. (In preparation.)

MICHAEL J. HARRISON, PH.D., is the Dean of Lyman Briggs College and Professor of Physics at Lyman Briggs College, Michigan State University, East Lansing, Michigan, USA. He was born in Chicago 1932, studied physics at the Harvard University, at the University of Leiden (The Netherlands) and at the University of Chicago (Ph.D. in physics in 1960). He is working in the field of theoretical physics and teaching physics at Michigan State University since 1961 (since 1968 as professor).

Main publications:

Magnetic Field Dependence of Ultrasonic Attenuation. Phys. Rev. Letters *1* (1958), 441.

Magnetic Attenuation of Sound in Semimetals: Longitudinal Waves. Phys. Rev. *119* (1960), 1260.

Collective Excitation of Degenerate Plasmas in Solids. J. Phys. Chem. Solids *23* (1962), 1079.

Resonant Excitation of Plasmons in Thin Films by Electromagnetic Waves (with A. R. Melnyk). Phys. Rev. Letters *21* (1968), 85.

Theory of Optical Excitation of Plasmons in Metals (with A. R. Melnyk). Phys. Rev. B *2* (1970), 835.

JOHANNES PETERS, DR.-ING. E.H., is scientific collaborator of Messerschmitt—Bölkow—Blohm GmbH, Ottobrunn bei München and of Gesellschaft für Mathematik und Datenverarbeitung mbH, Bonn. He was born in 1911 in Germany. After his high-school final examination he had to follow a practical professional career. He was occupied with engineering work at the German broadcast. In 1957 he changed over to an aircraft enterprise where he is still attached to as a consultant. In 1953 he became a lecturer on communication theory at the University of Hamburg. He was awarded with the title of Honorary Doctor by the Technical University of Hannover in 1966.

Main publications:

Wann gilt das Stabilitätskriterium nach Nyquist? A. E. Ü. *4* (1950), 17—22.

Einschwingvorgänge, Gegenkopplung, Stabilität. Springer-Verlag, Berlin—Göttingen—Heidelberg 1954.

Einführung in die allgemeine Informationstheorie. Springer-Verlag, Berlin—Heidelberg—New York 1967.

Information und Signal. In: K. Steinbuch und S. Moser: Philosophie und Kybernetik. Nymphenburger Verlagshandlung, München 1970, 163—173.

KENNETH GEORGE DENBIGH, PH.D., is the College Principal at Queen Elizabeth College, University of London, U.K. Born in Luton, England, in 1911, studied chemistry at Leeds University in 1929—1934; subsequently employed by Imperial Chemical Industries Ltd. and held university appointments at the University of Southampton and University of Cambridge. Subsequently appointed to professorships at University of Edinburgh (1955—1960) and at Imperial College, London (1960—1966). His main scientific interests are thermodynamics and theory of continuous processes.

Main publications:

The Thermodynamics of the Steady State. Methuen, 1951.

The Principles of Chemical Equilibrium. 3rd ed., Cambridge University Press, 1971.

Science, Industry and Social Policy. Oliver and Boyd, 1963.

Chemical Reactor Theory. 2nd ed., Cambridge University Press, 1971.

AVENIR IVANOVICH UYEMOV, PHDR., DSC., is professor of philosophy at the University of Odessa (USSR). He is interested in the philosophy of science and logic. Born in 1928 in Porenya (USSR), studied philosophy at the University of Moscow till 1949, received the title CSc. in 1952. Then, he was assistant professor of philosophy at the Pedagogical Institute in Ivanovsk, since 1964 professor in Odessa.

Main publications:

Things, Qualities and Meanings (in Russian). Mysl, Moscow 1963.

Analogy in Practice of Scientific Inquiry (in Russian). Nauka, Moscow 1970.

Logical Foundations in the Method of Modelling (in Russian). Nauka, Moscow 1972.

K. S. Trincher is scientific collaborator of the Institute of Biological Physics of the Academy of Sciences of the USSR in Moscow. Born in 1910 in Brno, Czechoslovakia, he studied medicine and physical chemistry in Vienna. Since 1934, he is living in the USSR. He worked originally in the All-Union Institute of Experimental Medicine in Moscow, and since 1957 has been at the Institute of Biological Physics of the Academy of Sciences.

Main publications:

Biology and Information—Elements of biological Thermodynamics. Nauka, Moscow 1965 and Consultants Bureau, New York 1965.

Spin—Lattice Interaction of Water and Protein Membranes in Cell Metabolism. Journ. theor. Biol. *34* (1972), 557.

Zur Thermodynamik der Konformationsänderungen des Hämoglobinmoleküls. Zeitschrift für Naturforschung *27* (1972), 5, 587.

Die informations-energetische Analyse der Konformationsarbeit des Hämoglobinmoleküls. Third International Conference on Chemical Thermodynamics, 3—7 September, 1973, Baden near Vienna, Austria.

The Non-Applicability of the Entropy Concept in Living Systems. In: Unity through Diversity — a Festschrift for Ludwig von Bertalanffy. Gordon and Breach Science Publishers, New York—London—Paris 1973.

Nicholas Georgescu-Roegen, M.A., D. Stat., is Distinguished Professor of Economics at the Vanderbilt University, Nashville, USA. He was born in 1906 in Constanza, Romania, studied mathematics at the University of Bucharest and statistics at the University of Paris. His post doctoral research was made under Karl Pearson at University College in London 1930—1932. He was professor of statistics at the University of Bucharest in 1932—1946, then lecturer and research associate at Harvard University in 1948—1949, Rockefeller visiting professor in Japan 1962—1963, Ford visiting professor in Brazil 1966, etc. He was Fellow of the Econometric Society (1950), Fellow of the International Institute of Sociology (1960), Distinguished Fellow of the American Economic Association (1971), Fellow of the American Academy of Arts and Sciences (1973) etc.

Main publications:

Activity Analysis of Production and Allocation (co-editor with T. C. Koopmans et al.). John Wiley & Sons, New York 1951.

Analytical Economics: Issues and Problems. Harvard University Press, Cambridge, Mass. 1966.

La science économique: Ses problèmes et ses difficultés. Dunod, Paris 1970.

The Entropy Law and the Economic Problem. University of Alabama, 1971 (pamphlet).

The Entropy Law and the Economic Process. Harvard University Press, Cambridge, Mass. 1971.

Analisi economica e processo economico. Sansoni, Firenze 1973.

Samuil Natanovich Braines, Dr.Med., DSc., is professor of medicine and chief of the Biocybernetic Laboratory of the Surgical Institute of the Medical Academy of the USSR in Moscow. Born in 1898 at Yuzovo (USSR), he studied medicine at the University of Krym till 1923 and received the title DSc. in 1936. He worked in the field of experimental biology and pathology, later in the field of biocybernetics.

Main publications:

Problems of Experimental Pathology (in Russian). Medgiz, Moscow 1950.

Problems of Neurocybernetics (in Russian). Medgiz, Moscow 1959.

Neurocybernetics (with V. B. Svetchinsky) (in Russian). Medgiz, Moscow 1962.

Problems of Neurocybernetics and Neurobionics (with V. B. Svetchinsky). Medicina, Moscow 1968.

Biological and Medical Cybernetics (with V. B. Svetchinsky) (in Russian). Medicina, Moscow 1971.

VLADISLAV BORISOVICH SVETCHINSKY, DR.MED.CSC., scientific collaborator of the Surgical Institute of the Medical Academy of the USSR in Moscow. Born in 1936 in Astrakhan (USSR), he studied at the Energetical Institute in Moscow till 1960, received degree CSc. in 1966. He works in the field of biocybernetics and technical cybernetics.

Main publications:
Co-author of 3 books with S. N. Braines (see Braines).

MICHAL STRÍŽENEC, PH.D., CSC is psychologist at the Institute of Experimental Psychology of the Slovak Academy of Sciences in Bratislava, Czechoslovakia. Born of Piešťany in 1932, he studied psychology at the Philosophical Faculty of Comenius University, Bratislava, graduating in 1955, and supporting his Candidate of Psychological Sciences thesis (CSc.) "Information Theory and Work Output" in 1960. His research work involved problems of cybernetics as applied in psychology, analysis of operator work in the chemical industry, and recently, psychological aspects of the man-computer system.

Main publications:

Psychology and Cybernetics (in Slovak). VSAV, Bratislava 1966.

Current Tendencies in Engineering Psychology (in Slovak). VSAV, Bratislava 1971.

Zur experimentellen Bestimmung der subjektiven Wahrscheinlichkeit. Ztschr. f. Psychologie *171* (1965), 377—382.

Quelques aspects méthodologiques de la recherche sur les processus intellectuels de l'opérateur. Travail Humain *31* (1968), 3—4, 269—282.

Prediction of parametre changes in a simulated control system (I, II). Studia psychologica *12* (1970), 214—226, 249—257.

JOSEF LINHART, PHDR., DRSC., Corresponding member of the Czechoslovak Academy of Sciences, Chairman of the Scientific Board of Pedagogy and Psychology, Professor of the Department of Psychology, Faculty of Philosophy, Charles University, Prague, Director of the Institute of Psychology of the Czechoslovak Academy of Sciences, Prague. Born in 1917 and educated at the Faculty of Philosophy, Charles University, Prague. In 1951 he submitted his thesis on "The influence of environment and education on the mental development of the child". In 1970 he was appointed Director of the Institute of Psychology of the Czechoslovak Academy of Sciences. In his scientific and research work he is concerned with the study of theoretical problems of general psychology with respect to psychology of learning and mental development, psychology of problem solving and creative activity. More recently he has been occupied with the relations of these psychological problems to cybernetics and information theory. The results of his experimental research based on his own theoretical conception were presented in approximately 200 scientific publications, both in Czechoslovakia and abroad, almost 50 being major works of synoptic nature.

Main publications:

The Psychological Problems of Theory of Learning (in Czech). Prague 1965.

The Psychology of Learning (in Czech). Prague 1967.

Chapters from Psychology of Learning, Thinking and Creative Activity (in Czech). Prague 1971.

Random Variations, Structure and Plan of Creative Activity (in Czech). Prague 1971.

The Process and Structure of Human Learning (in Russian). Moscow 1970. (In Czech; Prague 1972.)

STANISLAV VOHNÍK, graduate physicist, head of the Computer Laboratory, Institute of Psychology of the Czechoslovak Academy of Sciences, Prague. Born in 1940 in Prague and educated at the Faculty of Mathematics and Physics, Charles University, Prague. He held the post of Lecturer in the Department of Mathematics, Faculty of Electrical Engineering, University of Plzeň. He lectured on the theory of probability and on statistics with special regard to the applications in technical cybernetics. At present he is engaged on problems of representation of psychological functions.

Petr Horák, graduate mathematician, works at the Computer Laboratory, Institute of Psychology of the Czechoslovak Academy of Sciences, Prague. Born in 1949 at Karlovy Vary (Czechoslovakia). In the years 1967—1972 he read theoretical cybernetics at the Faculty of Mathematics and Physics, Charles University, Prague. In the Institute of Psychology he works in the field of application of mathematical logic in psychology.

Adolf Grünbaum, Ph.D., works as Andrew Mellon Professor of Philosophy and Director of the Center for Philosophy of Science, University of Pittsburgh, USA. Born 1923 in Cologne, Germany, he studied philosophy and physics and became B.A. in 1943 (Wesleyan University), M.S. in 1948 (Yale University) and Ph.D. in 1950 (Yale University). He was professor of philosophy at Le High University (1956—1960), since 1960 at the University of Pittsburgh. He received in 1958 the J. Walker Tomb Prize from Princeton University for publications on the philosophy of time. He was Vice-President of the American Association for the Advancement of Science (1963) and President of the Philosophy of Science Association (1965—1967, 1968—1970). He is a member of the American Philosophical Association, American Association for the Advancement of Science, etc. He published about 100 scientific articles and 4 books.

Main publications:

Modern Science and Zeno's Paradoxes. Wesleyan University Press, 1967. (Revised edition; Allen and Unwin, London 1968.)

Geometry and Chronometry in Philosophical Perspective. Univ. of Minnesota Press, 1968.

Philosophical Problems of Space and Time. Knopf, New York 1963. (In Russian: Progress, Moscow 1969. Second enlarged edition: Reidel, Dordrecht – Boston 1973).

Falsifiability and Rationality. University of Pittsburgh Press (forthcoming).

Arkadiy Dmitriyevich Ursul, PhDr., DSc., is scientific collaborator of the Philosophical Institute of the Academy of Sciences of the USSR in Moscow. His fields of interest are philosophy of science, cybernetics and cosmical research. Born in 1936 in Moldavia, he studied at the Aeronautical Institute in Moscow till 1959. He followed then the studies in philosophy, in 1964 received the title CSc. and in 1969 the title DSc. He is also professor of philosophy at the Pedagogical Institute in Moscow.

Main publications:

Capture of Cosmos (in Russian). Mysl, Moscow 1967.

Essence of Information (in Russian). Politizdat, Moscow 1968. (In German: Information. Dietz, Berlin 1970).

Information (in Russian). Nauka, Moscow 1971.

Reflection and Information (In Russian). Mysl, Moscow 1973.

Herbert Hörz, Prof., DrSc., Chairman of the Department "Philosophical Problems of the Development of Science", Academy of Sciences of the German Democratic Republic, Central Institute of Philosophy, Berlin. Born in 1933 in Stuttgart, he went to high school at Erfurt and he studied from 1952 to 1956 philosophy and physics in Jena and Berlin. He became a doctor of philosophy in 1960 and received another degree in 1962. He has been a lecturer at Humboldt University since 1963 and he became a professor of philosophical problems of natural sciences in 1965. He was a National Prize winner of the GDR in 1972. In 1973 he became a Corresponding member of the Academy of Science. His major publications deal with the interrelations of philosophy and physics.

Main publications:

Der dialektische Determinismus in Natur und Gesellschaft. Berlin 1962 (5. Aufl. 1973).

Atome – Kausalität – Quantensprünge. Berlin 1964.

Materie und Bewußtsein. Berlin 1965.

Werner Heisenberg und die Philosophie. Berlin 1966 (2. Aufl. 1968).

Physik und Weltanschauung. Berlin 1968.
Materiestruktur. Berlin 1971.

BENJAMIN GAL-OR, D.SC., is professor at the Department of Aeronautical Engineering, Technion, Israel Institute of Technology in Haifa (Israel). Born in Israel in 1933, received B.Sc. in 1959, M.Sc. in 1961 and D.Sc. in 1964 in Chemical Engineering from Technion-I.I.T. He was assistant professor at Johrs Hopkins University, then associate professor at the University of Pittsburgh. He works as professor at Technion since 1968. He received the Gold Medal (The Laskowitz Award for 1971) from the New York Academy of Sciences for original research in aerospace (for the development of the astrophysical thermodynamics). Organized and chaired the International Conference "A Critical Review of Relativistic and Classical Thermodynamics" in 1969, Pittsburgh, USA. His 75 publications deal with two-phase fluid-dynamics, shock waves, chemical reactions, thermodynamics and philosophy of science.

Main publications:

Gas-Liquid Dispersions. In: Advances in Chemical Engineering, Vol. 7. Academic Press, N.Y. 1968, 295—395.

A Critical Review of Thermodynamics (E. B. Stuart, B. Gal-Or, and A. J. Braindard, Editors). Mono Book Corporation, Baltimore 1970.

The Crisis about the Origin of Irreversibility and Time Anisotropy. Science *176* (1972), 11—17.

Modern Developments in Thermodynamics (Editor). John Wiley and Israel Universities Press, New York and Jerusalem 1974.

IRENA SZUMILEWICZ, docent hab., Doctor of philosophy, is working at the Institute of Sociology and Philosophy, Polish Academy of Sciences, Warsaw, Poland. She was born in Zambrów and lived in Warsaw till 1939. She received the master's degree in 1938 at Warsaw University, Department of Chemistry. In 1959 she received the doctor's degree and in 1964 got the docent's degree—both at Warsaw University, Department of Philosophy. She worked 16 years at Gdańsk University as a chairman of the philosophical department. Since 1973, she works at the Polish Academy of Sciences in Warsaw.

Main publications:

The Theory of Thermodynamical Death of the Universe (in Polish). PWN, Warszawa 1961.

The Direction of Time (in Polish). PWN, Warszawa 1964.

The Mechanicism of Ludwig Boltzmann and the Postulate of Microreduction (in Polish). TNT, Toruń 1969.

The Law of Nature According to Henri Poincaré (in Polish). In: Pojęcie prawa nauki a konwencjonalizm początku XX wieku. Zakład im. Ossolińskich, Wrocław—Warszawa—Kraków—Gdańsk 1972.

JIŘÍ ZEMAN, PHDR., CSC., is scientific collaborator of the Institute for Philosophy and Sociology of the Czechoslovak Academy of Sciences in Prague. His fields of interest are the philosophy of science and philosophical problems of cybernetics. Born in Tábor (Czechoslovakia) in 1926, he studied philosophy and psychology at Charles University in Prague, graduated in 1952 and received the title candidate of philosophical sciences (CSc.) at the Academy in 1960. He is working at the Academy of Sciences since 1954, since 1967 as leading collaborator.

Main publications:

Knowledge and Information (in Czech). NČSAV, Prague 1962. (In Russian; Progress, Moscow 1966).

Cybernetics and the Modern Science (in Czech). NPL, Prague 1964.

Cybernetics in Social Sciences (in Czech) (Editor). NČSAV, Prague 1965.

Time in Science and Philosophy (Editor). Academia—Elsevier, Prague—Amsterdam 1971.